COMPARISONS

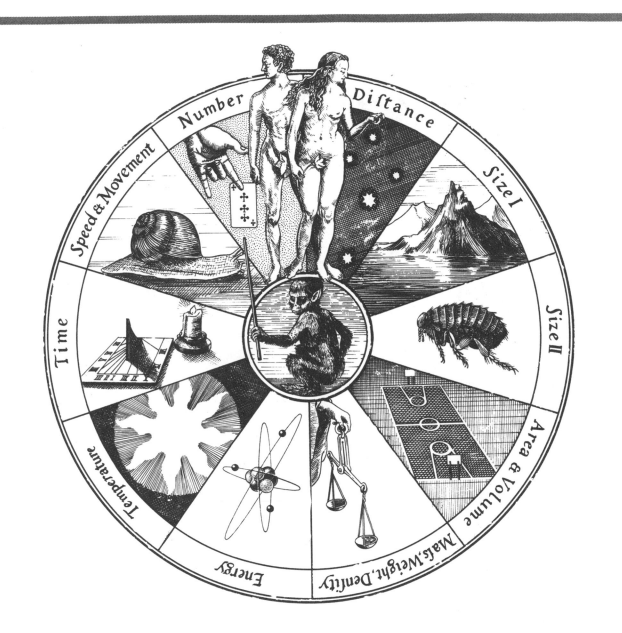

COMPARISONS

of distance, size, area, volume, mass, weight, density, energy, temperature, time, speed and number throughout the universe

by the Diagram Group

ST. MARTIN'S PRESS/NEW YORK

Library of Congress Cataloging in Publication Data
Diagram Group.
    Comparisons.

    1. Physical measurements—Handbooks, manuals, etc.
I. Title.
QC39.D5   530.8   80-14251
ISBN 0-312-15484-4

FOREWORD

"Comparisons" is an exciting visual guide to how man measures—and measures up to—his universe. By making comparisons between like and like and between like and unlike we can evaluate the different properties of the known world.

Man has devised many units and scales of measurement, and **"Comparisons"** brings these together in one handy reference volume, pointing out differences from system to system and providing methods of conversion.

Most readers are familiar with the excellent "Guinness Book of Records," which has for many years recorded the extremes of the human and physical worlds. **"Comparisons"** displays and relates not only extremes, but also presents what we encounter in everyday experience.

To make this wealth of information easier to understand and remember, the artists and editors have created hundreds of illustrations and diagrams, all accompanied by concise, explanatory captions. This technique makes it possible to see at a glance and to appreciate fully comparisons that are not readily grasped from photographs or prose alone.

The originality of **"Comparisons"** will, we hope, both fascinate and inform, and also stimulate every reader to make further comparisons of his own.

The Diagram Group

Managing editor Ruth Midgley

Research editors Hope Cohen; Norma Jack

Contributors Jeff Cann; Cornelius Cardew; Michael Carter;
Maureen Cartwright; Marion Casey; Sam Elder;
David Lambert; Gail Lawther; Linda Proud;
Bernard Moore; Angela Royston; Marita Westberg

Art editor Richard Hummerstone

Artists Steven Clark; Mark Evans; Brian Hewson; Susan Kinsey;
Janos Marffy; Graham Rosewarne, Kathleen McDougall

Art assistants Steve Clifton; Richard Colville; Neil Connell;
Steve Hollingshead; Richard Prideaux; Ray Stevens

Picture researcher Enid Moore

Indexer Mary Ling

Acknowledgments
The authors and publishers wish to extend their warmest thanks to the many individuals, institutions and companies who have responded with great patience and generosity to numerous research enquiries. Special thanks are due to the following:

Amateur Athletic Association; American Embassy, London;
Australia House, London; Robert Barry; Damaris Batchelor;
British Airways; British Museum; British Olympic
Association; British Petroleum Co. Ltd; British Rail;
Cottie G. Burland; Cunard Steamship Co. Ltd; Department
of Energy, UK; Ford Motor Co. Ltd; Greater London Council;
Guinness Superlatives Ltd; Christopher Hand; Hovercraft
Developments Ltd; Institute of Geological Sciences, London;
International Glaciological Society; Intourist; Library of the
Zoological Society of London; Lloyds Register of Shipping;
London Transport; Military Archive and Research Services;
Museum of Mankind, London; National Maritime Museum,
London; National Motor Museum, Beaulieu, Hants;
Natural History Museum, London; Railway Magazine;
Royal Geographical Society, London; Royal Greenwich
Observatory; School of Oriental and African Studies,
London; Science Museum, London; Martin Suggett; United
Nations London Information Centre; Claudio Vita-Finzi

Dedication
This book is dedicated to:
Tom McCormack, who thought of the idea;
Len and Elkie Shatzkin, who introduced Tom to the Diagram Group;
Ruth Midgley, who had to work harder than she had ever done before; and
Patricia Robertson, who brought up four children single-handed while Bruce Robertson was too busy to come home

CONTENTS

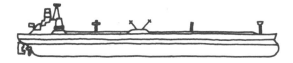

CHAPTER 1

Confirming the length of the rood, an old linear measure, by lining up and measuring the feet of the first 16 men out of church —from Jacob Köbel's *Geometrey von künstlichen Feldtmessen* of 1598 (Science Museum, London).

DISTANCE

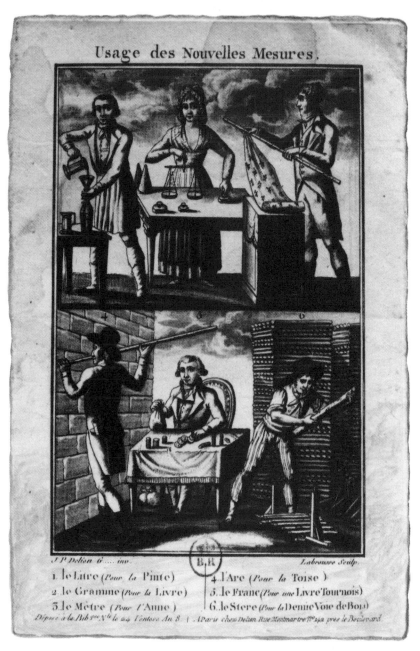

A contemporary French print illustrating the use of the new metric measures introduced in the wake of the French Revolution of 1789 (Photo: Bulloz).

MEASURING DISTANCE

The width of a finger, the length of a foot, the distance covered in a stride, and the length of a furrow ploughed by a horse are ingenious linear measurements from earlier times. Today, demands for greater precision and standardization are resulting in a much wider acceptance of the scientifically based metric system.

US customary/imperial units of linear measurement	
12 inches (in)	= 1 foot (ft)
3 feet	= 1 yard (yd)
1760 yards	= 1 mile (mi)
Metric units of linear measurement	
10 millimeters (mm)	= 1 centimeter (cm)
100 centimeters	= 1 meter (m)
1000 meters	= 1 kilometer (km)

The body rules *below*
Illustrated are two ancient measuring systems based on the human body.
Egyptian measurements
A Digit, one finger width
B Palm (= four digits)
C Hand (= five digits)
D Cubit, elbow to finger tips (= 28 digits, 20.6in)

Roman measurements
E Foot, length of one foot (subdivided into 12 *unciae*, hence our inches)
F Pace (= 5 feet), of which 1000 made up the Roman mile (*mille passus*)

Basic units *above*
Included here are the most commonly used units of linear measurement, both US/imperial and metric. Given in brackets are the standard abbreviations used in this book. (For additional tables, see pp. 14, 16, 18.)

Ready measures *below*
These common objects can be used as convenient measures for US/imperial and metric distances. They are shown here real size together with a US/imperial and metric scale.
a Key 2in
b Cent ¾in
c Paper clip 1¼in
d Shirt button 1cm
e Paper match 4cm

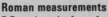

It's a good thing that some of our great poets of former days used the league as a unit of measurement. Imagine Tennyson's famous poem beginning "A mile and a half, a mile and a half, a mile and a half onwards," instead of "Half a league, half a league . . ."

The ell was once widely used for measuring cloth, but lengths varied: the French ell was 54in, the English 45in and the Flemish only 27in.

12 douzièmes = 1 line
4 lines = 1 barleycorn
3 barleycorns = 1 inch

5½ yards = 1 rod
4 rods = 1 chain
10 chains = 1 furlong
8 furlongs = 1 mile
3 miles = 1 league

Unusual US/imperial units *left* Originating in a largely agricultural society, most of the imperial units of length given in this table have now virtually died out.

Conversion tables *below* These tables can be used to convert US/imperial distances (inches, feet, yards and miles) into metric (centimeters, meters and kilometers), and vice versa. For example, to convert 5in into centimeters, find 5 in the center column of the first table and then read the figure opposite it in the right-hand column (12.700cm). To find the inch equivalent of 5cm, find 5 in the center column and then read the corresponding left-hand column figure (1.9685in).

in		cm	ft		m	yd		m	mi		km
0.3937	1	2.5400	3.2808	1	0.3048	1.0936	1	0.9144	0.6214	1	1.6093
0.7874	2	5.0800	6.5617	2	0.6096	2.1872	2	1.8288	1.2427	2	3.2187
1.1811	3	7.6200	9.8425	3	0.9144	3.2808	3	2.7432	1.8641	3	4.8280
1.5748	4	10.160	13.123	4	1.2192	4.3744	4	3.6576	2.4855	4	6.4374
1.9685	5	12.700	16.404	5	1.5240	5.4680	5	4.5720	3.1069	5	8.0467
2.3622	6	15.240	19.685	6	1.8288	6.5617	6	5.4864	3.7282	6	9.6560
2.7559	7	17.780	22.966	7	2.1336	7.6553	7	6.4008	4.3496	7	11.265
3.1496	8	20.320	26.247	8	2.4384	8.7489	8	7.3152	4.9710	8	12.875
3.5433	9	22.860	29.528	9	2.7432	9.8425	9	8.2296	5.5923	9	14.484
5.9055	15	38.100	49.213	15	4.5720	16.404	15	13.716	9.3206	15	24.140
9.8425	25	63.500	82.021	25	7.6200	27.340	25	22.860	15.534	25	40.233
13.779	35	88.900	114.83	35	10.668	38.276	35	32.004	21.748	35	56.327
17.716	45	114.30	147.64	45	13.716	49.212	45	41.148	27.962	45	72.420
21.654	55	139.70	180.45	55	16.764	60.149	55	50.292	34.175	55	88.514
25.591	65	165.10	213.25	65	19.812	71.085	65	59.436	40.389	65	104.61
29.528	75	190.50	246.06	75	22.860	82.021	75	68.580	46.603	75	120.70
33.465	85	215.90	278.87	85	25.908	92.957	85	77.724	52.817	85	136.79
37.402	95	241.30	311.68	95	28.956	103.89	95	86.868	59.030	95	152.89

Critical distances *left, below* Some examples from the world of sport:
1 Soccer penalty spot to goal mouth center, 12yd
2 Bowling lane, 60ft long
3 Pitcher to batter at baseball, 60ft 6in
4 Cricket pitch, distance between wickets, 22yd

© DIAGRAM

"Full fathom five thy father lies," sings Ariel in Shakespeare's *Tempest*. He meant that Ferdinand's father had been drowned and was lying at a depth of 30ft; a fathom is a mainly nautical unit equal to 6ft.

Common rule *right* Shown real size is part of a rule with both US/imperial and metric divisions.

THE MICROSCOPIC WORLD

Our vision of the world is literally restricted, for with our eyes alone we can see only objects that are above a certain size. We can, for instance, see the dot above an "i" but without a microscope a single grain of most types of pollen is invisible. Here we look at measurements in the microscopic world.

Explanation of scales
below Starting with a real-size illustration of a section of a rule marked with both inches and centimeters, we have drawn a series of scales, each of which represents a 10-fold magnification of a tenth of the previous one.

Measurements here—and on other pages where we have similar scales—are expressed in metric units, since the decimal character of the metric system makes it ideal when using factors of 10. On each scale we state the distance that 1cm represents.

A
B
C
D
E

1cm : 1mm (10^{-3}m)
1cm : 0.1mm (10^{-4}m)
1cm : 0.01mm (10^{-5}m)
1cm : 1μm (10^{-6}m)
1cm : 0.1μm (10^{-7}m)

Meter	m		
Decimeter	dm	10^{-1}m	0.1m
Centimeter	cm	10^{-2}m	0.01m
Millimeter	mm	10^{-3}m	0.001m
Micrometer	μm	10^{-6}m	0.000 001m
Nanometer	nm	10^{-9}m	0.000 000 001m
Picometer	pm	10^{-12}m	0.000 000 000 001m
Femtometer	fm	10^{-15}m	0.000 000 000 000 001m
Attometer	am	10^{-18}m	0.000 000 000 000 000 001m

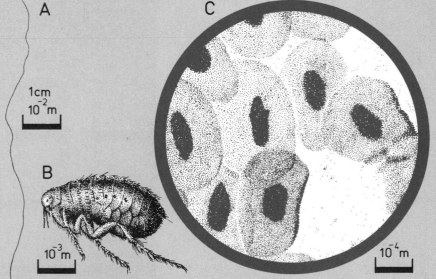

A

1cm
10^{-2}m

B

10^{-3}m

C

10^{-4}m

Small metric units
above The table lists metric units for measuring small and microscopic distances. We start with the meter, the base unit of the International System of Units (SI), defined as 1,650,763.73 wavelengths in vacuum of the orange-red line of the spectrum of krypton-86. This is followed by the names of smaller units, together with their abbreviations, and meter equivalents expressed first as powers of 10 and then as decimals.

An attometer, the smallest unit of linear measurement, is to a shirt button what a shirt button is to a planetary system with a diameter 8.6 times that of the solar system.

If each page in a 1000 page book were only 1 micrometer (micron) thick, then all the pages together would measure only 1 millimeter, half the thickness of the gray line below.

If magnified according to a scale where 1 centimeter represents 1 picometer (10^{-12} cm), a small raindrop (diameter 1.4 millimeters) would be as large as the Sun (diameter approximately 1,400,000 kilometers).

Small made large
bottom Each illustration has been drawn to a different one of our scales, shown by a letter. In this way we are able to see how the application of progressively larger scales brings smaller and smaller objects into view.

A The first illustration in our sequence is of a human hair, real size.
B Next we have a flea, which has been drawn to scale B, i.e. magnified 10 times. Each centimeter of our flea is equivalent to one millimeter in real life.

C The red blood cells of a frog are drawn to scale C, appearing in our illustration as they would when seen under a microscope magnifying them 100 times. Each centimeter of the enlarged image is equivalent to one tenth of a millimeter in real life.

D Next we have a selection of viruses drawn to scale D. They appear in our illustration as they would under an electron microscope magnifying them 100,000 times. Each centimeter represents one tenth of a micrometer.

E Here we see iron atoms (large dots) and sulfur atoms (small dots) as they appear in a crystal of marcasite (iron sulphide) that has been magnified 10 million times (scale E). Each centimeter is equivalent to one nanometer.

F At the right edge of the page we show part of a sodium atom that has been drawn to scale F. This represents a magnification of 1000 million. On this scale, one centimeter of our drawing is equivalent to one hundreth of a nanometer.

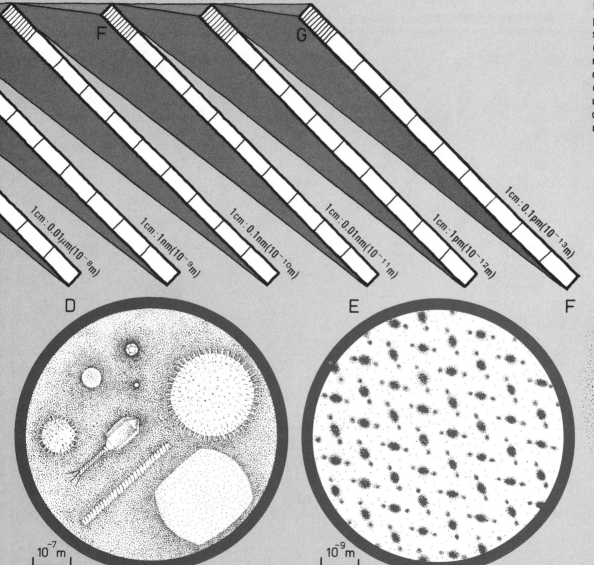

1cm : 0.01μm(10^{-8}m)

1cm : 1nm(10^{-9}m)

1cm : 0.1nm(10^{-10}m)

1cm : 0.01nm(10^{-11}m)

1cm : 1pm(10^{-12}m)

1cm : 0.1pm(10^{-13}m)

10^{-7} m

10^{-9} m

10^{-11} m

If we were to draw our solar system to a scale where one femtometer of our illustration represented one centimeter in reality, it would be possible to fit more than 400 such illustrations on a single page of this book.

G The dot in the box represents the nucleus of a hydrogen atom drawn to scale G, i.e. magnified 100,000 million times. One centimeter in the box is equivalent to one tenth of a picometer.

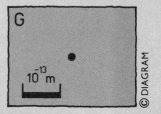

10^{-13} m

OUR WORLD AND BEYOND 1

On the preceding two pages we used progressively larger scales to look down into the microscopic world. Here and on the next two pages we use progressively smaller scales to allow us to visualize distances in the world and out into space. In doing so, detail is lost, but bigger and bigger distances are brought into view.

Explanation of scales
below As on the previous two pages, we start with part of a centimeter/inch rule and then follow this with a series of scales. But here our scales become smaller, each new scale reducing the preceding scale to one tenth.

Large made small
bottom Our sequence of drawings shows how by using progressively smaller scales we encompass bigger and bigger distances. **A** We start with a picture of a junction in New York, drawn to scale A where 1cm is equivalent to 10m.

A   B   C   D   E

1cm:10cm   1cm:1m   1cm:10m   1cm:100m   1cm:1km   1cm:10km

Meter	m		
Dekameter	dam	10m	10m
Hectometer	hm	10^2m	100m
Kilometer	km	10^3m	1000m
Myriameter	mym	10^4m	10,000m
Megameter	Mm	10^6m	1,000,000m
Gigameter	Gm	10^9m	1,000,000,000m
Terameter	Tm	10^{12}m	1,000,000,000,000m

Large metric units *above*
The table begins with a meter, the base SI unit of length (see p.14 for definition), and then gives larger units, with abbreviations, and meter equivalents given first as powers of 10 and then numerically.

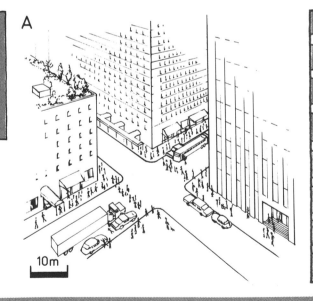

A

10m

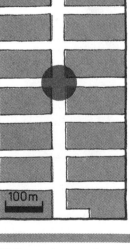

B

100m

The world's longest river, the Nile, would be the length of an average shoelace (66.7cm) if drawn to our scale D (1cm:100km).

If drawn to a scale where 1cm is equivalent to 10m (our scale A) this book would be no bigger than the dot on this i.

If drawn to our scale D, where 1cm represents 100km, an average raindrop (diameter 2mm) would have the dimensions of a smallish atom (diameter 0.000 000 2mm).

B Our New York junction is here drawn to scale B, where 1cm is equivalent to 100m in reality. We now see the junction in the context of the streets around it, as though looking at it from above, or locating it on a city street plan.

C This map is drawn to scale C, where 1cm is equivalent to 10km. We can no longer see the junction or the pattern of the city streets. Instead we have a detailed view of the geographical location of New York City and its neighbors.

D Long Island appears as a prominent feature on this map where 1cm represents 100km (scale D). The area shown on map C is on this map indicated by a box. Around it we see parts of New York State, Philadelphia, Massachusetts and Connecticut.

E Our drawing of part of a globe has been made to scale E, where 1cm is equivalent to 1000km. We can no longer identify Long Island, but can now see the whole of the East coast of North America, Central America and much of South America as well.

F When drawn to scale F, where 1cm is equivalent to 100,000km, Earth and Moon appear as tiny dots. Our scale diagram does, however, allow us to visualize the distance between them—3.8cm as drawn, or approximately 380,000km in reality.

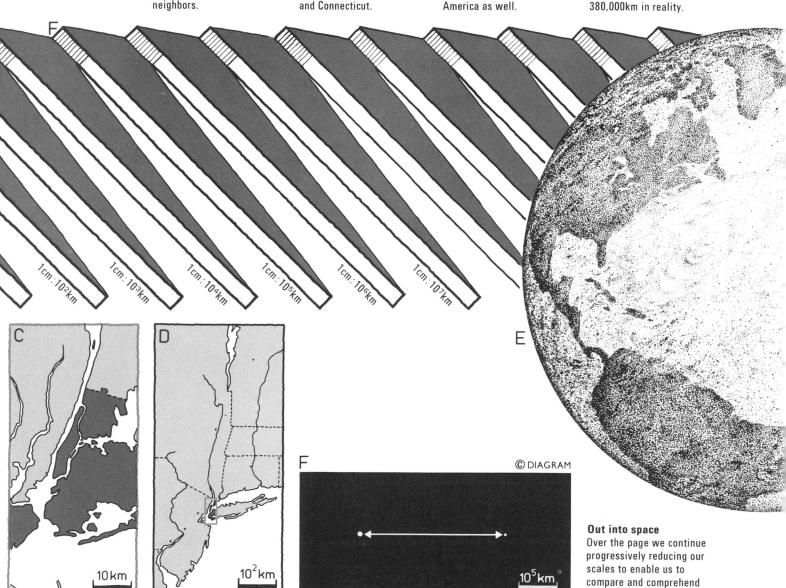

F

$1cm:10^2km$ $1cm:10^3km$ $1cm:10^4km$ $1cm:10^5km$ $1cm:10^6km$ $1cm:10^7km$

C

D

F

© DIAGRAM

E

10km

10^2km

10^5km

Out into space
Over the page we continue progressively reducing our scales to enable us to compare and comprehend the vast distances of space.

Distances at sea are measured in nautical miles. The international nautical mile is equal to 1852m (6076ft); the UK nautical mile, based on the original definition, is equal to the mean length of one minute of longitude (6080ft or 1.152 statute miles).

OUR WORLD AND BEYOND 2

Out into space
Here we continue the series of scales begun on the previous two pages. As before, each scale reduces the preceding scale to one tenth. Colored bars are used to show the change over to light years as our unit of measurement.

G This diagram of part of the solar system is drawn to scale G, with 1cm equivalent to 10^8km. Shown are the orbits of the inner planets, the distance between Sun and Earth(**1**), the asteroid belt (**2**), the orbit of Jupiter (**3**) and of Halley's comet (**4**).

H Scale H is 100 times smaller than scale G; 1cm now represents 10^{10}km. On this scale the solar system appears quite insignificant. Shown are the orbits of the outer planets around the Sun (with Pluto a mean 5900 million km away), and the orbit of Halley's comet.

I On scale I, 1cm represents 10^{12}km. The long scale bar on our diagram shows 1 ly drawn to this scale. Also shown is the distance between the Sun and its nearest stellar neighbor, Proxima Centauri, some 40,207,125 million km or 4.25 ly away.

G H I J K

1cm : 10^8km
1cm : 10^9km
1cm : 10^{10}km
1cm : 10^{11}km
1cm : 10^{12}km
1cm : 1 ly
1cm : 10 ly
1cm : 10^2 ly

Astronomical unit (au)
1 au = 149,600,000km = 93,000,000mi
Light year (ly)
1 ly = 9,460,500,000,000km = 5,878,000,000,000mi
Parsec (pc)
1 pc = 30,857,200,000,000km = 19,174,000,000,000mi
63,240au = 1 ly
206,265au = 1 pc
3.262 ly = 1 pc

Units for space *above*
The table lists standard abbreviations and equivalents of units used in the measurement of astronomical distances. These units are further defined as follows. An astronomical unit is the mean distance between the Earth and the Sun.
A light year is the distance traveled in one year by electro-magnetic waves in vacuo.
A parsec is the distance at which a base line of 1 astronomical unit in length subtends an angle of 1 second.

G

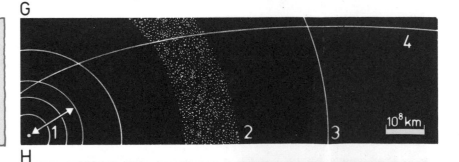

10^8 km

H

10^{10} km

Using light years as a unit may save us from writing a long string of zeros but it is not always easy to appreciate the great distances involved. For example, the radio galaxy 3C–295 is over 26,000,000,000,000,000,000,000 miles away!

J When drawn to scale J where 1cm represents 10 ly, the Sun and Proxima Centauri, here indicated by arrows, appear very close together, as indeed they are in galactic terms.

K Here the central part of our galaxy is drawn to scale K, where 1cm is equivalent to 10^4ly. The diameter of the entire galaxy is about 100,000ly, or 10cm on this scale.

L Our galaxy, indicated by an arrow, is shown with other galaxies in our local group in this illustration to scale L, where 1cm is equivalent to 10^6ly.

M This illustration of M104 (popularly called the Sombrero galaxy, in the Virgo cluster of galaxies) has not been drawn to scale, but the distance between M104 and Earth, some 41 million ly, occurs within scale M, where 1cm is equivalent to 10^7ly.

N Indicated on the illustration by an arrow is the radio galaxy 3C-295, one of the most distant galaxies ever photographed. Its distance from Earth, some 5000 million ly, occurs within scale N, where 1cm is equivalent to 10^9ly.

Toward infinity
Our scales continue off the page toward infinity. As yet, however, the greatest distance claimed for an object detected in space is 15,600 million ly, which could be measured on the scale where 1cm represents 10^{10}ly.

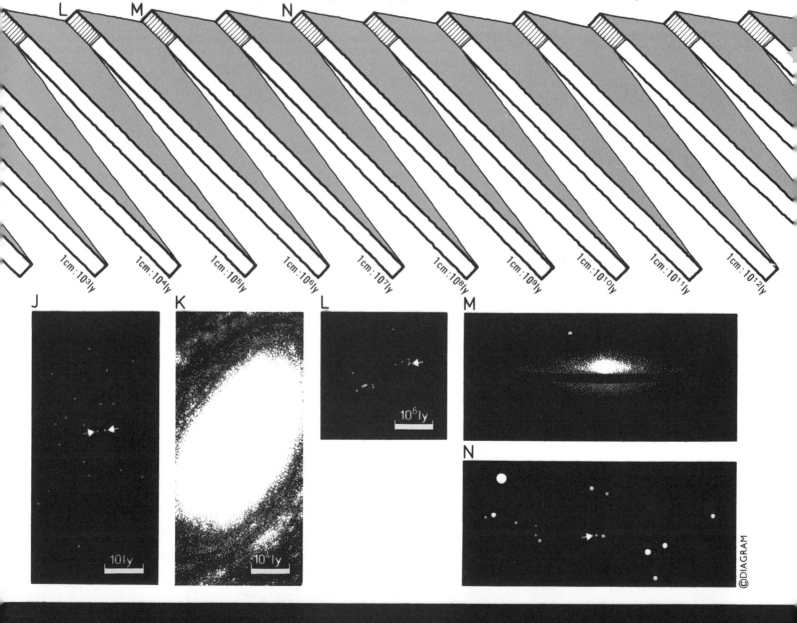

1cm:10^3ly   1cm:10^4ly   1cm:10^5ly   1cm:10^6ly   1cm:10^7ly   1cm:10^8ly   1cm:10^9ly   1cm:10^{10}ly   1cm:10^{11}ly   1cm:10^{12}ly

J — 10 ly

K — 10^4 ly

L — 10^6ly

N

©DIAGRAM

1ly

DISTANCES RUN AND WALKED

Distance presents man with a variety of sporting challenges. Here we compare the distances run and walked in Olympic events, where the object is to cover a prescribed distance as fast as possible. Also compared are distance endurance records, where the aim is to travel as far as possible in a particular manner.

Around the world *right*
Some of man's more unusual distance endurance records are here shown to scale as if they had been achieved along the Equator, 40,075km (24,900mi) long.

1 Walking on hands, 1400km (871mi), equal to one twenty−ninth of the length of the Equator.
2 Swimming, 2938km (1826mi), one fourteenth of the length of the Equator.
3 Running, 8,224km (5,110mi), equal to one fifth of the Equator.
4 Backward walking, 12,875km (8000mi), one third of the Equator.
5 Walking, 48,000km (29,825mi), equal to one and a fifth times the length of the Equator.
6 Cycling, 160,934km (100,000mi) in one tour, just over four times the length of the Equator.

Olympic distances *right*
Listed are running/walking distances, given in meters in keeping with International Olympic Committee practice, plus approximate US/imperial distance equivalents, and whether there is an event for men, women or both.

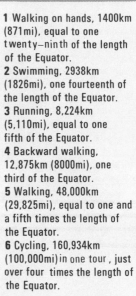

a 100m (110yd) men/women	**h** 800m (880yd) men/women
b 100m (110yd) hurdles women	**i** 1500m (1640yd) men/women
c 110m (120yd) hurdles men	**j** 4 x 400m (440yd) relay men/women
d 200m (220yd) men/women	**k** 3000m (3280yd) steeplechase men
e 400m (440yd) men/women	**l** 5000m (3.1mi) men
f 400m (440yd) hurdles men	**m** 10,000m (6.2mi) men
g 4 x 100m (110yd) relay men/women	**n** 20km (12.4mi) walk men
	o 42.195km (26mi 385yd) marathon men
	p 50km (31mi) walk men

Running for a train
right The diagram shows the Olympic sprint and hurdle race distances to scale with a passenger train whose cars are a typical 30m (100ft) long. The longest sprint race, the 400m, is slightly longer than 13 such cars.

Racing through New York
right Middle and long distance Olympic running and walking events are here shown to scale with a map of Manhattan Island, New York. The 10,000m would take you from Battery Park (**A**) at the southern tip of the island as far as the middle of Central Park (**B**). The longest Olympic road walk, the 50km (31mi) walk is approximately equal to walking from Battery Park (**A**) to Inwood Hill Park (**C**) and back again.

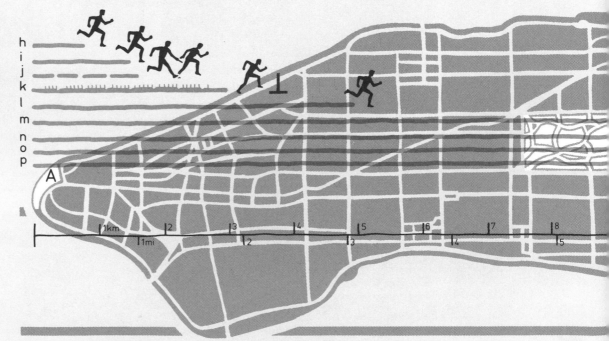

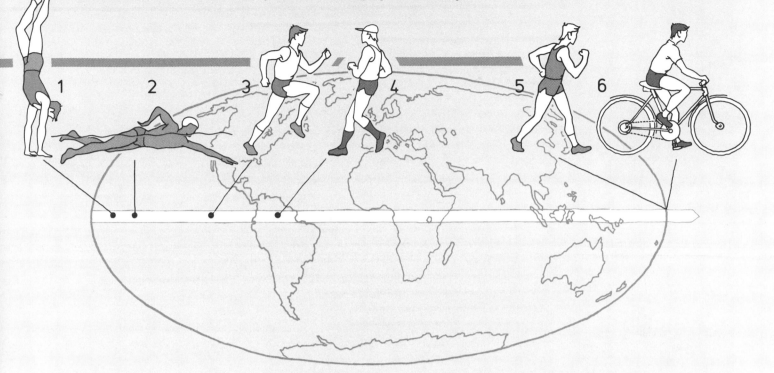

A ringed Arctic tern found in Fremantle, Western Australia, had flown half way around the world—12,000 miles, from a White Sea coast reserve in the USSR. The human long-distance walking record is roughly 2½ times as far.

Olympic records
For a comparison of running and walking records see our chapter on Speed.

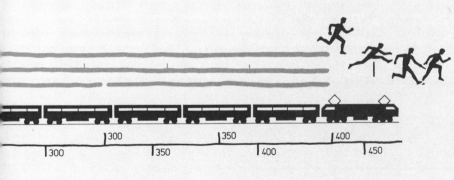

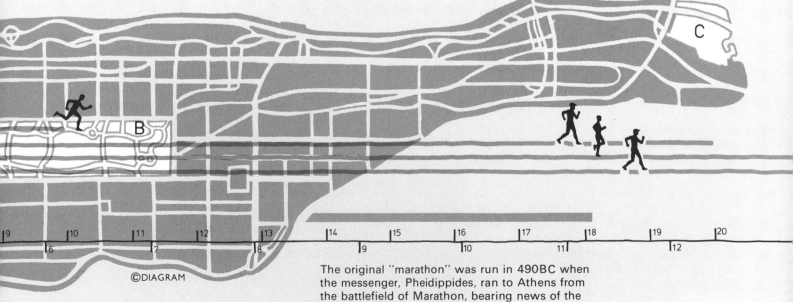

©DIAGRAM

The original "marathon" was run in 490BC when the messenger, Pheidippides, ran to Athens from the battlefield of Marathon, bearing news of the Athenian army's great victory over the Persians.

DISTANCES JUMPED

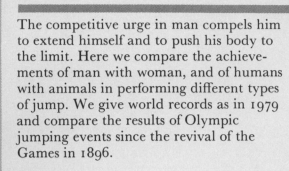

The competitive urge in man compels him to extend himself and to push his body to the limit. Here we compare the achievements of man with woman, and of humans with animals in performing different types of jump. We give world records as in 1979 and compare the results of Olympic jumping events since the revival of the Games in 1896.

How far can we jump? *below* Human long jump performances are here compared with those of three animals known for their jumping ability— the flea, the frog and the kangaroo. An average man is about 9in taller than a red kangaroo, but the world long jump record is less than three-fourths of the distance that the kangaroo can cover with ease. Relative to its size, the performance of the flea is even more impressive: to equal its achievement man would have to leap about 400yd!

A   B

		15	16	17	18	19	
			5				6
					1948		

Jumping Jacks *right*
1 Amateur standing long jump record, 11ft 11¾in, clearing two average men lying head to foot.
2 World long jump record, clearing 5 men.
3 World triple jump record, 58ft 8½in, clearing 10 men with ease.

High divers *left*
Professional divers at Acapulco, Mexico dive into water from rocks 118ft high, equal to a dive from the roof of an 11-story building!

The world long jump record for men is equal to the length of two Ford Mustangs or two Jaguar E-types parked fender to fender.

To better the world pole vault record it will be necessary to exceed the height of three average men and a 5-year-old boy standing one on top of another's shoulders.

Long jump results
A Flea 13in
B Frog 17ft 6¾in
C Women's world record, Vilma Bardauskiene (USSR) 23ft 3¼in
D Men's world record, Robert Beamon (USA) 29ft 2½in
E Red kangaroo 40+ft

©DIAGRAM

1896 | 1900 | 1908 | 1912 | 1928 | 1936 | 1960 | 1968
22 | 23 | 24 | 26 | 27 | 28 | 29 | ft
7 | 8 | 9 | m
1956 | 1978 | 1952 | 1968

Farther and farther
above The diagram allows us to compare record long jump performances from 1896 to the present. Dates when records were set are shown here on a distance scale. In 1896 E. Clark (USA) jumped 20ft 10in— a distance equaled by the women's Olympic record in 1956 (E. Krzesinska, Poland). In 1968, R. Beamon (USA) increased the men's Olympic record by an amazing 2ft 7in. This is, however, 1¾in less than the amount by which the 1896 record was bettered in 1900.

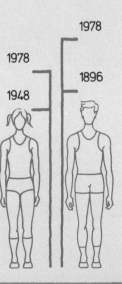

1978
1978 | 1896
1948

How high can we jump?
above Shown are the heights reached by the women's and men's world high jump records, the world pole vault record (men), and the official high jump record for horses. Kangaroos have cleared an unofficial 10½ft.

Higher and higher
left This diagram allows us to compare current world high jump records with the heights attained when the high jump became a modern Olympic event for men and for women. Also in scale are a man and woman of average height.

Heights cleared
1 Women's world high jump record, Sara Simeoni (Italy) 6ft 7in
2 Men's world high jump record, Vladimir Yashchenko (USSR) 7ft 8in
3 Horse, official FEI high jump record 8ft 1¾in
4 World pole vault record. David Roberts (USA) 18ft 8¼in

If the holder of the men's world high jump record had made his jump on Mercury where gravity is weaker, he would have cleared 20ft 8in, exceeding the world pole vault record by almost 2ft.

HEIGHTS AND DEPTHS REACHED

Man has achieved truly amazing feats of vertical mobility. He has climbed to the top of Earth's highest mountain, even without the use of breathing equipment; in a spacecraft he has orbited the Moon. Without breathing equipment he has dived down 282ft; in a bathyscaphe he has been to the bottom of the deepest ocean trench.

Reaching the heights
Plotted here against a logarithmic scale and listed in the table *right* are details of altitude records achieved since the first successful balloon trials in 1783. Within the first year the record was boosted from only 80ft to an incredible 9000ft. In 1923 the general altitude record was for the first time taken by an aircraft. From 1933 until 1951 this record was again held by balloons, but since then first a rocket and rocket planes and then spacecraft have been highest of all.

Heights achieved by man	
1 Hot air balloon, tethered (1783)	80ft
2 Hot air balloon, tethered (1783)	324ft
3 Hot air balloon (1783)	3000ft
4 Hydrogen balloon (1783)	9000ft
5 Hydrogen balloon (1803)	20,000ft
6 Coal gas balloon (1837)	25,000ft
7 Coal gas balloon (1875)	27,950ft
8 Aircraft (1923)	36,565ft
9 Aircraft (1930)	43,166ft
10 Hydrogen balloon (1933)	61,237ft
11 Helium balloon (1935)	72,395ft
12 Skyrocket (1951)	79,600ft
13 Rocket plane (1954)	93,000ft
14 Rocket plane (1956)	23.9mi
15 Rocket plane (1961)	203.2mi
16 Spacecraft (1966)	850.7mi
17 Spacecraft (1968)	234,473mi
18 Spacecraft (1970)	248,655mi

Life at high altitudes
Also included on our graph *below* are the height of Mount Everest, and the highest recorded altitudes for various forms of life.
a Mount Everest 29,002ft
b Alpine chough, highest recorded bird, 26,900ft
c Non-flowering plant 20,833ft
d Flowering plant 20,130ft
e Yak 20,000ft
f Highest inhabited human dwelling 19,700ft
g Bacteria 135,000ft

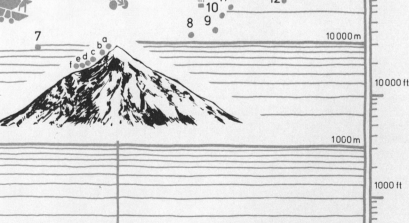

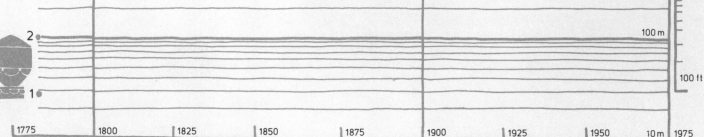

Depths achieved by man

*Women's record
**Died in the attempt
***Simulated dive
†First scuba record

A) Breath-held dives
1 Georghios (1913) c200ft
2 Mayol (1966) 198ft
3 Croft (1967) 212ft 6in
4 Treleani (1967) 147ft 6in*
5 Croft (1967) 217ft 6in
6 Mayol (1968) 231ft
7 Croft (1968) 240ft
8 Maiorca (1971) 250ft
9 Mayol (1973) 282ft

B) Dives using air
1 Lambert (1885) 162ft
2 Greek and Swedish divers (1904) 190ft
3 Damant (1906) 210ft
4 Drellifsak (1914) 274ft
5 Crilley/Nielson/Loughman (1915) 304ft
6 Hilton (1932) 344ft
7 Dumas (1947) 307ft†
8 Farques (1947) 396ft**
9 Root (1947) 400ft**
10 Clarke-Samazen (1954) 350ft
11 Troutt (1964) 320ft*
12 Watts/Johnson (1966) 355ft
13 Watts/Munns (1967) 380ft
14 Giesler (1967) 325ft*
15 Gruener/Watson (1968) 437ft

C) Dives using gas mixtures
1 Nohl (1937) 420ft
2 Metzger/Conger (1941) 440ft
3 Zetterstrom (1945) 528ft**
4 Bollard/Soper (1948) 450ft
5 Johnson (1949) 550ft
6 Wookey (1956) 600ft
7 Keller/Macleish (1961) 728ft
8 Keller/**Small (1962) 1000ft
9 USN Aquanauts (1968) 1025ft***
10 Deckman (1968) 1100ft***
11 Brauer/Veyrunes (1968) 1197ft***
12 Bevan/Sharphouse (1970) 1500ft***
13 Chemin/Gauret (1972) 2001ft***

D) Dives in machines
1 Steel sphere (1865) c245ft
2 Diving bell (1889) c830ft
3 Hydrostat (1911) c1650ft
4 Bathysphere (1930) 1426ft
5 Bathysphere (1932) 2200ft
6 Bathysphere (1934) 2510ft
7 Bathysphere (1934) 3028ft
8 Benthoscope (1949) 4500ft
9 Bathyscaphe (1953) 5085ft
10 Bathyscaphe (1953) 6890ft
11 Bathyscaphe (1953) 10,335ft
12 Bathyscaphe (1954) 13,287ft
13 Bathyscaphe (1959) 18,600ft
14 Bathyscaphe (1960) 24,000ft
15 Bathyscaphe (1960) 35,802ft

A) Breath-held dives
Dives of around 120ft have long been made by divers collecting oysters and sponges. Georghios' 1913 record for a dive without breathing equipment lasted until the challenges of Mayol and Croft in the late 1960s.

B) Dives using air
The development of scuba (self-contained underwater breathing apparatus) has led to the establishment of impressive new underwater diving records. The record scuba dive using air is 437ft—over 150ft deeper than the breath-held record.

C) Dives using gas mixtures
Breathing gas mixtures allows divers to go even deeper than when air is breathed. In 1962 divers established a record of 1000ft after release from a diving bell; since then simulated dives have doubled the record.

D) Dives in machines
Man's deepest dives have been made in a variety of machines developed for the purpose. The latest of these, the bathyscaphe, has taken the record dive in a machine down from 4500ft to 35,802ft, on the seabed in the Marianas Trench.

Deep-sea life *right*
Also listed and shown here are the greatest depths at which various animals have been recorded. Divers at 35,802ft saw what may have been a fish.

Deep-sea life
a Seal 1968ft
b Sperm whale 3720ft
c Sponge 18,500ft
d *Bassogigas profundissimus* 23,230ft
e Amphipod 32,119ft

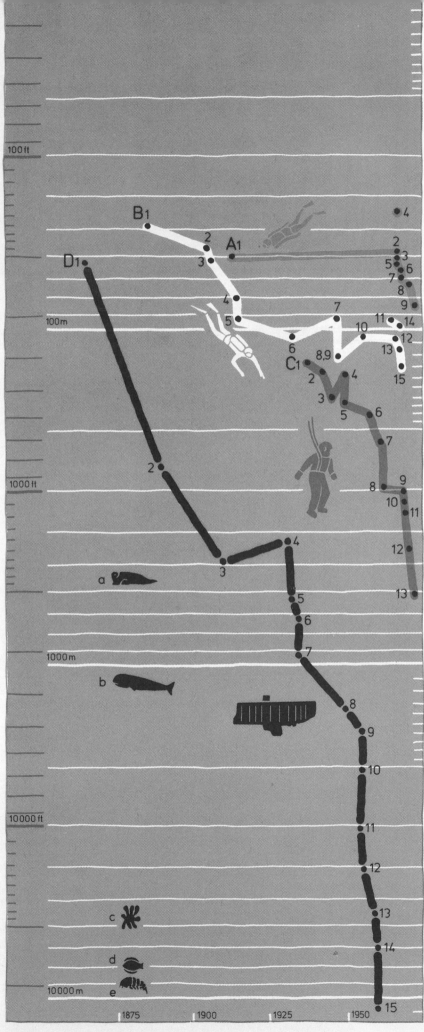

DISTANCES FROM PLACE TO PLACE

The shortest theoretical surface route between two places on Earth can never exceed 12,451 miles, that is one half of Earth's circumference. Among major cities, Wellington/Paris (11,791 miles), Wellington/London (11,682 miles), and Rio de Janeiro/Tokyo (11,535 miles) most nearly approach this maximum.

Shortest distances *below*
The table gives the shortest distances between selected cities around the world. These "great circle" distances (see explanation *right*) are given here in miles (colored band) and kilometers.

	Berlin, Germany	Bombay, India	Cape Town, South Africa	Darwin, Australia	London, England	Los Angeles, USA	Mexico City, Mexico	Moscow, USSR	New York, USA	Paris, France	Peking, China	Port Said, Egypt	Quebec, Canada
Berlin, Germany		3910 / 6292	5977 / 9619	8036 / 12,932	574 / 924	5782 / 9305	6037 / 9715	996 / 1603	3961 / 6374	542 / 872	4567 / 7350	1747 / 2811	3583 / 5766
Bombay, India	3910 / 6292		5134 / 8262	4503 / 7247	4462 / 7181	8701 / 14,003	9722 / 15,646	3131 / 5039	7794 / 12,543	4359 / 7015	2964 / 4770	2659 / 4279	7371 / 11,862
Cape Town, South Africa	5977 / 9619	5134 / 8262		6947 / 11,180	6005 / 9664	9969 / 16,043	8511 / 13,697	6294 / 10,129	7801 / 12,554	5841 / 9400	8045 / 12,947	4590 / 7387	7857 / 12,644
Darwin, Australia	8036 / 12,932	4503 / 7247	6947 / 11,180		8598 / 13,837	7835 / 12,609	9081 / 14,614	7046 / 11,339	9959 / 16,027	8575 / 13,800	3728 / 5999	7159 / 11,521	9724 / 15,649
London, England	574 / 924	4462 / 7181	6005 / 9664	8598 / 13,837		5439 / 8753	5541 / 8917	1549 / 2493	3459 / 5567	213 / 343	5054 / 8133	2154 / 3466	3101 / 4990
Los Angeles, USA	5782 / 9305	8701 / 14,003	9969 / 16,043	7835 / 12,609	5439 / 8753		1542 / 2482	6068 / 9765	2451 / 3944	5601 / 9014	6250 / 10,058	7528 / 12,115	2579 / 4150
Mexico City, Mexico	6037 / 9715	9722 / 15,646	8511 / 13,697	9081 / 14,614	5541 / 8917	1542 / 2482		6688 / 10,763	2085 / 3355	5706 / 9183	7733 / 12,445	7671 / 12,345	2454 / 3949
Moscow, USSR	996 / 1603	3131 / 5039	6294 / 10,129	7046 / 11,339	1549 / 2493	6068 / 9765	6688 / 10,763		4662 / 7503	1541 / 2480	3597 / 5789	1710 / 2752	4242 / 6827
New York, USA	3961 / 6374	7794 / 12,543	7801 / 12,554	9959 / 16,027	3459 / 5567	2451 / 3944	2085 / 3355	4662 / 7503		3622 / 5829	6823 / 10,980	5590 / 8996	439 / 706
Paris, France	542 / 872	4359 / 7015	5841 / 9400	8575 / 13,800	213 / 343	5601 / 9014	5706 / 9183	1541 / 2480	3622 / 5829		5101 / 8209	1975 / 3178	3235 / 5206
Peking, China	4567 / 7350	2964 / 4770	8045 / 12,947	3728 / 5999	5054 / 8133	6250 / 10,058	7733 / 12,445	3597 / 5789	6823 / 10,980	5101 / 8209		4584 / 7377	6423 / 10,337
Port Said, Egypt	1747 / 2811	2659 / 4279	4590 / 7387	7159 / 11,521	2154 / 3466	7528 / 12,115	7671 / 12,345	1710 / 2752	5590 / 8996	1975 / 3178	4584 / 7377		5250 / 8449
Quebec, Canada	3583 / 5766	7371 / 11,862	7857 / 12,644	9724 / 15,649	3101 / 4990	2579 / 4150	2454 / 3949	4242 / 6827	439 / 706	3235 / 5206	6423 / 10,337	5250 / 8449	
Rio de Janeiro, Brazil	6144 / 9888	8257 / 13,288	3769 / 6065	9960 / 16,029	5772 / 9289	6296 / 10,132	4770 / 7676	7179 / 11,553	4820 / 7757	5703 / 9178	10,768 / 17,329	6244 / 10,048	5125 / 8248
Rome, Italy	734 / 1181	3843 / 6185	5249 / 8447	8190 / 13,180	887 / 1427	6326 / 10,180	6353 / 10,224	1474 / 2372	4273 / 6877	682 / 1098	5047 / 8122	1317 / 2119	3943 / 6345
Tokyo, Japan	5538 / 8912	4188 / 6740	9071 / 14,598	3367 / 5419	5938 / 9556	5470 / 8803	7035 / 11,321	4650 / 7483	6735 / 10,839	6033 / 9709	1307 / 2103	5842 / 9402	6417 / 10,327
Wellington, New Zealand	11,265 / 18,129	7677 / 12,355	7019 / 11,296	3310 / 5327	11,682 / 18,800	6714 / 10,805	6899 / 11,103	10,279 / 16,542	8946 / 14,397	11,791 / 18,975	6698 / 10,779	10,249 / 16,494	9228 / 14,851

Long and winding roads
right Great circle distances are of only limited use when planning a journey by car. Natural features and cities make routes far from direct.

		1 Great circle distance	2 Extra distance by road	% extra	
a	Berlin/Paris	542mi (872km)	124mi (200km)	23%	a
b	Berlin/Rome	734mi (1181km)	224mi (360km)	31%	b
c	London/Rome	887mi (1427km)	292mi (470km)	33%	c
d	London/Moscow	1549mi (2493km)	294mi (473km)	19%	d
e	Los Angeles/Mexico City	1542mi (2482km)	475mi (764km)	31%	e
f	Los Angeles/New York	2451mi (3944km)	464mi (747km)	19%	f

The world's longest continuous frontier, between the USA and Canada, is 3987 miles long—more than 500 miles longer than the great circle distance between New York and London.

The world's shortest frontier, between Spain and Gibraltar, is only 1672 yards long—less than half the width, at 42nd Street, of Manhattan Island, New York.

A great circle *right*
Any two points on the Earth's surface can be connected by a great circle line, which traces the shortest surface route between them. Great circle lines are formed by slicing a globe in half through its center.

Getting it straight *right*
The use of a map "projection" allows us to transfer the curved surface of the Earth onto a flat map. This process inevitably causes some distortion, different projections distorting different features while keeping others reasonably accurate. Included here for comparison are maps drawn to three different projections. Marked on each of them is the great circle line between New York and Moscow; although this line represents the shortest actual distance between these cities, it appears as an arc not as a straight line even on map C, where the distortion of this area of the globe is fairly slight. Our maps are details from the following projections:
A Mercator
B Polyconic
C Polar azimuthal equidistant

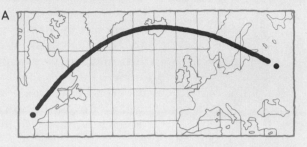
A

B

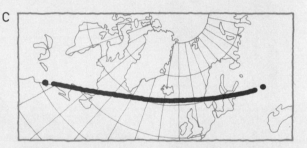
C

	Rio de Janeiro, Brazil	Rome, Italy	Tokyo, Japan	Wellington, New Zealand	
6144 / 9888	734 / 1181	5538 / 8912	11,265 / 18,129	Berlin, Germany
8257 / 13,288	3843 / 6185	4188 / 6740	7677 / 12,355	Bombay, India
3769 / 6065	5249 / 8447	9071 / 14,598	7019 / 11,296	Cape Town, South Africa
9960 / 16,029	8190 / 13,180	3367 / 5419	3310 / 5327	Darwin, Australia
5772 / 9289	887 / 1427	5938 / 9556	11,682 / 18,800	London, England
6296 / 10,132	6326 / 10,180	5470 / 8803	6714 / 10,805	Los Angeles, USA
4770 / 7676	6353 / 10,224	7035 / 11,321	6899 / 11,103	Mexico City, Mexico
7179 / 11,553	1474 / 2372	4650 / 7483	10,279 / 16,542	Moscow, USSR
4820 / 7757	4273 / 6877	6735 / 10,839	8946 / 14,397	New York, USA
5703 / 9178	682 / 1098	6033 / 9709	11,791 / 18,975	Paris, France
10,768 / 17,329	5047 / 8122	1307 / 2103	6698 / 10,779	Peking, China
6244 / 10,048	1317 / 2119	5842 / 9402	10,249 / 16,494	Port Said, Egypt
5125 / 8248	3943 / 6345	6417 / 10,327	9228 / 14,851	Quebec, Canada
	5684 / 9147	11,535 / 18,563	7349 / 11,827	Rio de Janeiro, Brazil
5684 / 9147		6124 / 9855	11,524 / 18,546	Rome, Italy
11,535 / 18,563	6124 / 9855		5760 / 9270	Tokyo, Japan
7349 / 11,827	11,524 / 18,546	5760 / 9270		Wellington, New Zealand

Antipodal points *left*
A city's "antipodal point" is the place on Earth most distant from it. On a globe this can be found by projecting a line through the center and out the other side. In practice, however, "antipodes" are found by calculation.

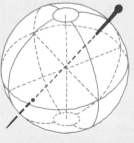

A world apart *below*
Shown on this Mercator map are six cities and their antipodal points.
1 Los Angeles
2 New York
3 Rio de Janeiro
4 London
5 Moscow
6 Tokyo

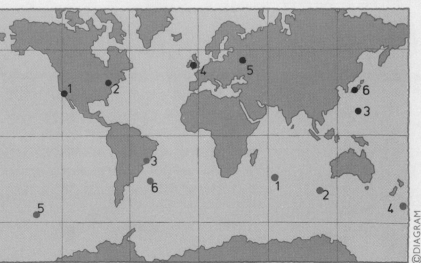

©DIAGRAM

THE SOLAR SYSTEM

Perpetual movement of objects in the solar system means that distances between them are always changing. As we show here, differences in relative positions can be immense. For example, the difference between Pluto's maximum and minimum distances from the Sun is almost 20 times the mean distance from Earth to Sun.

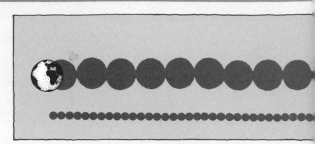

Earth's atmosphere *left*
This scale illustration allows us to compare the extent of the various layers occurring within Earth's atmosphere.
A Troposphere, upper limit about 5mi (8km) above ground at the Poles, 8mi (13km) at the Equator.

B Stratosphere, up to a limit of about 30mi (50km).
C Mesosphere, to about 50mi (80km) above ground.
D Thermosphere, up to an average in the region of 220mi (350km).
E Exosphere, with traces of hydrogen up to about 5000mi (8000km).

Worlds away *above*
The mean distance from Earth to Moon (center to center) is 238,840mi (384,365km). This is 30.1 times the diameter of Earth and 110.7 times that of the Moon. Minimum (**A**) and maximum (**B**) distances are also shown.

A down to Earth scale
below Mean distances of the planets from the Sun are compared here with relative distances across the USA. If the Sun were at New York, Pluto (**9**) would be at Los Angeles, and Jupiter (**5**) near Pittsburgh.

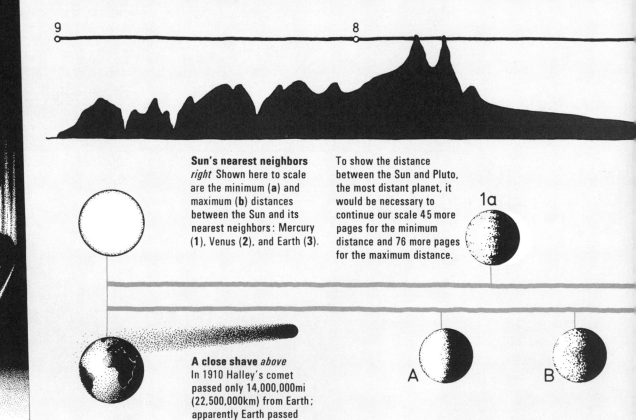

Sun's nearest neighbors *right* Shown here to scale are the minimum (**a**) and maximum (**b**) distances between the Sun and its nearest neighbors: Mercury (**1**), Venus (**2**), and Earth (**3**).

To show the distance between the Sun and Pluto, the most distant planet, it would be necessary to continue our scale 45 more pages for the minimum distance and 76 more pages for the maximum distance.

A close shave *above*
In 1910 Halley's comet passed only 14,000,000mi (22,500,000km) from Earth; apparently Earth passed through the comet's tail.

When closest to us, Venus, our nearest planetary neighbor, is 105 times more distant than our Moon. Neptune, however, never comes nearer than 11,208 times the distance from Earth to Moon.

Planetary orbits *right*
Since their orbits around the Sun are elliptical, planets have a minimum or "perihelion" (**A**) and a maximum or "aphelion" (**B**) distance from the Sun.

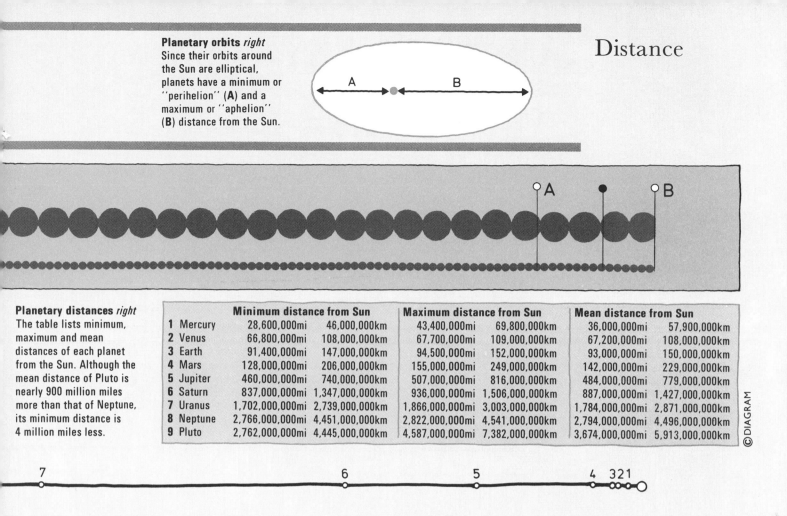

Planetary distances *right*
The table lists minimum, maximum and mean distances of each planet from the Sun. Although the mean distance of Pluto is nearly 900 million miles more than that of Neptune, its minimum distance is 4 million miles less.

		Minimum distance from Sun		Maximum distance from Sun		Mean distance from Sun	
1	Mercury	28,600,000mi	46,000,000km	43,400,000mi	69,800,000km	36,000,000mi	57,900,000km
2	Venus	66,800,000mi	108,000,000km	67,700,000mi	109,000,000km	67,200,000mi	108,000,000km
3	Earth	91,400,000mi	147,000,000km	94,500,000mi	152,000,000km	93,000,000mi	150,000,000km
4	Mars	128,000,000mi	206,000,000km	155,000,000mi	249,000,000km	142,000,000mi	229,000,000km
5	Jupiter	460,000,000mi	740,000,000km	507,000,000mi	816,000,000km	484,000,000mi	779,000,000km
6	Saturn	837,000,000mi	1,347,000,000km	936,000,000mi	1,506,000,000km	887,000,000mi	1,427,000,000km
7	Uranus	1,702,000,000mi	2,739,000,000km	1,866,000,000mi	3,003,000,000km	1,784,000,000mi	2,871,000,000km
8	Neptune	2,766,000,000mi	4,451,000,000km	2,822,000,000mi	4,541,000,000km	2,794,000,000mi	4,496,000,000km
9	Pluto	2,762,000,000mi	4,445,000,000km	4,587,000,000mi	7,382,000,000km	3,674,000,000mi	5,913,000,000km

©DIAGRAM

Halley's comet *below*
Indicated on our scale is the minimum distance of Halley's comet from the Sun: 55,000,000mi (89,000,000km).

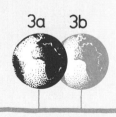

Orbit of Halley's comet
Our scale would have to extend another 54 pages to show this comet's greatest distance from the Sun: 3,281,400,000mi (5,280,800,000km).

Earth's nearest neighbors
left Shown to the same scale as the distances between the Sun and its neighbors (*above*) are the shortest distances between Earth and its nearest neighbors: Venus (**A**), Mars (**B**), and Mercury (**C**).

		Mean distance from Earth	
A	Venus	25,000,000mi	40,200,000km
B	Mars	35,000,000mi	56,300,000km
C	Mercury	50,000,000mi	80,500,000km
D	Jupiter	367,000,000mi	591,000,000km
E	Saturn	744,000,000mi	1,197,000,000km
F	Uranus	1,606,000,000mi	2,585,000,000km
G	Pluto	2,670,000,000mi	4,297,000,000km
H	Neptune	2,677,000,000mi	4,308,000,000km

Distances from Earth
left The table shows how close the other planets come to Earth. If we were to continue the scale showing the distances of Earth's nearest neighbors, we would find both Pluto and Neptune after 44 pages.

The mean distance from the Sun to Pluto, the outermost planet, is over 100 times greater than the mean distance between the Sun and Mercury, the innermost planet.

CHAPTER 2

A 19th-century depiction of the Tower of Babel. Excavations and written sources suggest that the tower of the Bible story was the ziggurat of the temple of Marduk at Babylon, believed to have stood 300ft tall on a square base with sides 300ft long (Mansell Collection).

God using dividers to measure the Universe, which is depicted as a series of concentric circle in this 17th-century engraving.

SIZE: NATURAL FEATURES AND MAN'S CONSTRUCTIONS

The Flatiron Building, New York City. At the time of its construction, in 1902, it was the tallest building in the world (Museum of the City of New York).

NATURAL FEATURES 1

We are often impressed by great heights in nature, but from our own relatively diminutive standpoint it is often difficult to appreciate the true scale involved. By comparing natural features with each other and also with possibly more familiar manmade objects it is possible to bring even Mt Everest into closer perspective.

Angel over the Empire State *below*
The world's highest waterfall, the Angel in Venezuela, is over twice the height of the Empire State Building (1472ft with mast). Shown here to scale with the Empire State are the heights of the world's 10 highest waterfalls; in each case the height is that of the total drop, which in some instances is made up of several smaller falls.

3212ft Angel, Venezuela
3110ft Tugela, S Africa

2625ft Utigôrd, Norway
2540ft Mongefossen, Norway
2425ft Yosemite, USA
2154ft Østre Mardøla Foss, Norway
2120ft Tyssestrengane, Norway
2000ft Kukenaom, Venezuela
1904ft Sutherland, NZ
1841ft Kjellfossen, Norway

Himalayan giants *right*
Some 25 peaks in the Himalayas exceed 20,000ft. We show the eight highest.
A Everest 29,002ft
B Godwin Austen 28,250ft
C Kanchenjunga 28,208ft
D Makalu 27,824ft
E Dhaulagiri 26,810ft
F Nanga Parbat 26,660ft
G Annapurna 26,505ft
H Gasherbrum I 26,470ft
Highest volcanoes
Of these, Cotopaxi is active, Llullaillaco is quiescent, the others are believed extinct. All are in S America except for Kilimanjaro and Elbrus.
I Aconcagua 22,834ft
J Llullaillaco 22,057ft
K Chimborazo 20,560ft
M Cotopaxi 19,344ft
N Kilimanjaro 19,340ft
O Antisana 18,713ft
P Citlaltepetl 18,700ft
Q Elbrus 18,480ft
Top peaks by continent
A Everest (Asia) 29,002ft
I Aconcagua (S Am) 22,834ft
L McKinley (N Am) 20,320ft
N Kilimanjaro (Afr) 19,340ft
Q Elbrus (Eur) 18,480ft
R Vinson Massif (Antarctica) 16,863ft
S Wilhelm (Oceania) 15,400ft
Other high spots
T Cook (NZ) 12,349ft
U Kosciusko (Australia) 7316ft
V Ben Nevis (UK) 4406ft
W Highest point in the Netherlands 321ft

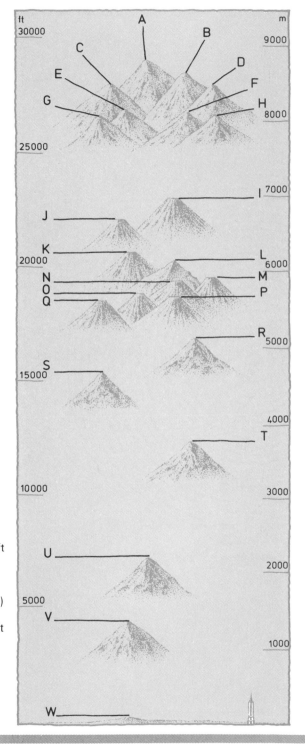

According to the Bible, Noah's ark came to rest on Mount Ararat after the water subsided. Since this mountain is 16,946ft high, it really must have been quite some flood!

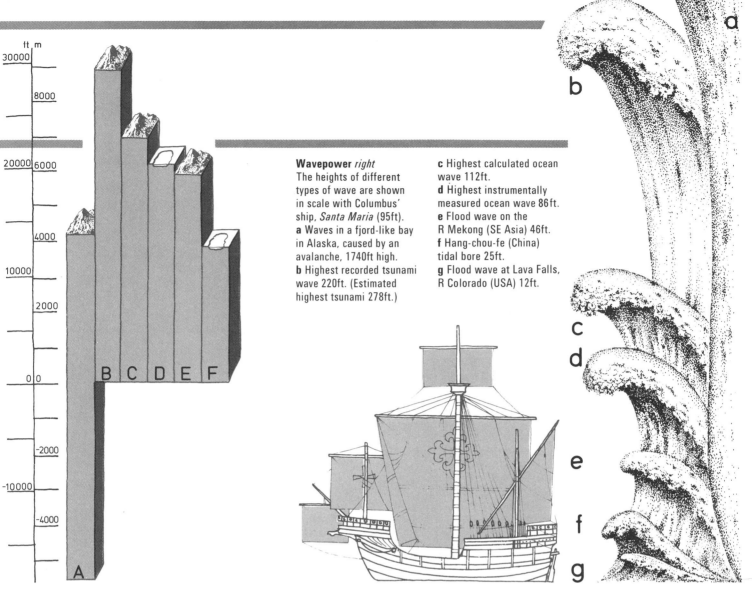

Wavepower *right*
The heights of different types of wave are shown in scale with Columbus' ship, *Santa Maria* (95ft).
a Waves in a fjord-like bay in Alaska, caused by an avalanche, 1740ft high.
b Highest recorded tsunami wave 220ft. (Estimated highest tsunami 278ft.)

c Highest calculated ocean wave 112ft.
d Highest instrumentally measured ocean wave 86ft.
e Flood wave on the R Mekong (SE Asia) 46ft.
f Hang-chou-fe (China) tidal bore 25ft.
g Flood wave at Lava Falls, R Colorado (USA) 12ft.

High features *above*
A Mauna Kea (Hawaii) is 4448ft taller than Mt Everest, but only 13,796ft of its total 33,476ft are above sea level.
B Everest, the world's highest peak, 29,002ft
C Aconcagua, highest extinct volcano, 22,834ft.
D World's highest lake, un-named (Tibet), 20,230ft.
E Cotopaxi, highest active volcano, 19,344ft.
F Titicaca (Peru/Bolivia), world's highest steam-navigated lake, 12,506ft.

State of elevation *right*
The highest point in the Netherlands is only 321ft above sea level, less than one-quarter the height of the Empire State Building (1472ft with mast).

Skyscraper rock *left*
Balls Pyramid, a rock pinnacle near Lord Howe Island, off the E coast of Australia, is 1843ft tall, making it 371ft taller than the Empire State Building (1472ft with mast).

©DIAGRAM

The deepest land depression, below the ice in Marie Byrd Land, Antarctica, dips 8100ft below sea level. The deepest known point in the ocean, in the Marianas Trench in the Pacific Ocean, is over four times as deep—36,198ft below sea level, or over 7000ft deeper than Mount Everest is high.

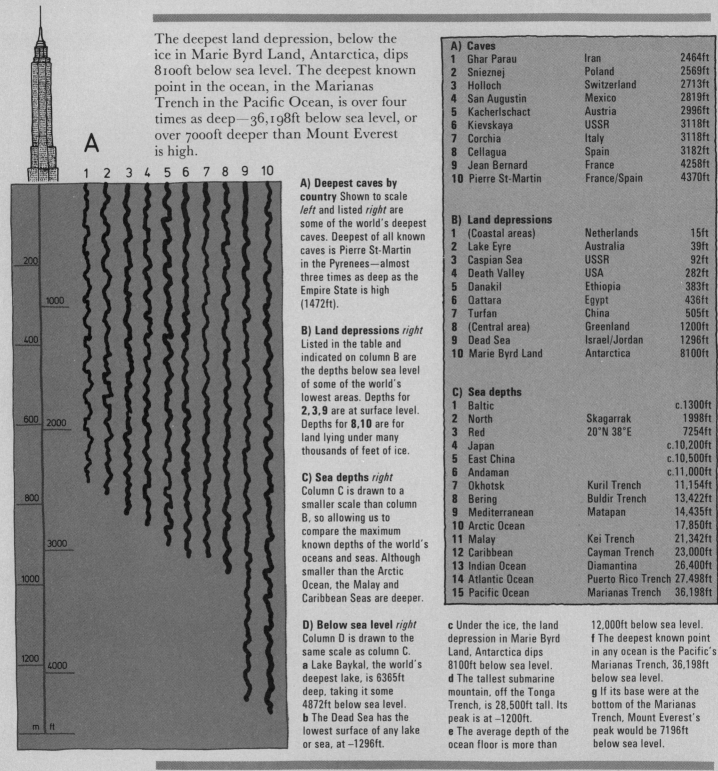

A) Deepest caves by country Shown to scale *left* and listed *right* are some of the world's deepest caves. Deepest of all known caves is Pierre St-Martin in the Pyrenees—almost three times as deep as the Empire State is high (1472ft).

B) Land depressions *right* Listed in the table and indicated on column B are the depths below sea level of some of the world's lowest areas. Depths for **2,3,9** are at surface level. Depths for **8,10** are for land lying under many thousands of feet of ice.

C) Sea depths *right* Column C is drawn to a smaller scale than column B, so allowing us to compare the maximum known depths of the world's oceans and seas. Although smaller than the Arctic Ocean, the Malay and Caribbean Seas are deeper.

D) Below sea level *right* Column D is drawn to the same scale as column C.
a Lake Baykal, the world's deepest lake, is 6365ft deep, taking it some 4872ft below sea level.
b The Dead Sea has the lowest surface of any lake or sea, at –1296ft.

c Under the ice, the land depression in Marie Byrd Land, Antarctica dips 8100ft below sea level.
d The tallest submarine mountain, off the Tonga Trench, is 28,500ft tall. Its peak is at –1200ft.
e The average depth of the ocean floor is more than

12,000ft below sea level.
f The deepest known point in any ocean is the Pacific's Marianas Trench, 36,198ft below sea level.
g If its base were at the bottom of the Marianas Trench, Mount Everest's peak would be 7196ft below sea level.

A) Caves		
1 Ghar Parau	Iran	2464ft
2 Snieznej	Poland	2569ft
3 Holloch	Switzerland	2713ft
4 San Augustin	Mexico	2819ft
5 Kacherlschact	Austria	2996ft
6 Kievskaya	USSR	3118ft
7 Corchia	Italy	3118ft
8 Cellagua	Spain	3182ft
9 Jean Bernard	France	4258ft
10 Pierre St-Martin	France/Spain	4370ft

B) Land depressions		
1 (Coastal areas)	Netherlands	15ft
2 Lake Eyre	Australia	39ft
3 Caspian Sea	USSR	92ft
4 Death Valley	USA	282ft
5 Danakil	Ethiopia	383ft
6 Qattara	Egypt	436ft
7 Turfan	China	505ft
8 (Central area)	Greenland	1200ft
9 Dead Sea	Israel/Jordan	1296ft
10 Marie Byrd Land	Antarctica	8100ft

C) Sea depths		
1 Baltic		c.1300ft
2 North	Skagarrak	1998ft
3 Red	20°N 38°E	7254ft
4 Japan		c.10,200ft
5 East China		c.10,500ft
6 Andaman		c.11,000ft
7 Okhotsk	Kuril Trench	11,154ft
8 Bering	Buldir Trench	13,422ft
9 Mediterranean	Matapan	14,435ft
10 Arctic Ocean		17,850ft
11 Malay	Kei Trench	21,342ft
12 Caribbean	Cayman Trench	23,000ft
13 Indian Ocean	Diamantina	26,400ft
14 Atlantic Ocean	Puerto Rico Trench	27,498ft
15 Pacific Ocean	Marianas Trench	36,198ft

Traveling at the speed of the fastest elevator in the Empire State Building it would take just over 30 minutes to reach the bottom of the Marianas Trench.

Under-statement *right*
The Empire State (1472ft with mast) is shown on the land mass below the ice in Marie Byrd Land. 4½ more Empire States are needed to reach sea level.

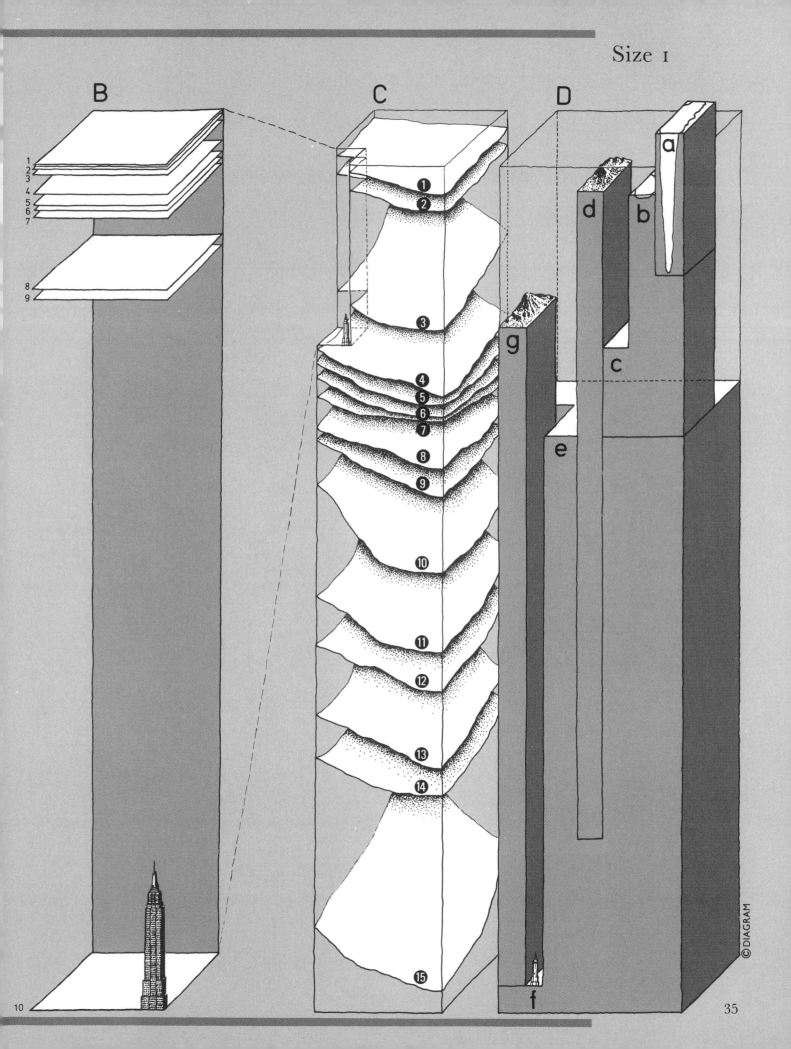

Counted among nature's longest features are mountain ranges and rivers longer than the width of North America, a glacier that stretches farther than the width of Switzerland, and a cave system that exceeds the length of Long Island, NY. Compared here are the world's longest rivers and mountain ranges, drawn to scale across North America to illustrate their enormous length; also shown is a selection of glaciers and cave systems. The longest mountain range (4500mi) is over 14 times the length of the longest glacier (Lambert-Fisher, 320mi) and more than 24 times the length of the longest cave system (Flint Ridge, 181.4mi).

Longest mountain ranges
Listed in the table *right* and drawn to scale *far right* are the world's longest mountain ranges. Although the Himalayas boast the world's highest mountains, this range ranks only third in the world in terms of length.

Longest rivers
right, far right Here we compare the world's longest rivers. Of the top 10, five are in Asia, two in Africa, two in N America and one in S America. Europe's longest river, the Volga, ranks only 16th.

Equator to Arctic *left*
The world's longest river, the Nile (4132mi), begins at the Equator and flows North to the Nile delta. If it began at the delta and flowed North along a Great Circle line, it would reach to within 100 miles of the North Pole.

Rivers of ice
Listed *right* and shown to scale with a map of Switzerland *left* are the world's longest glaciers. All but two of those listed are in Antarctica. Novaya Zemlya is much the longest glacier in the Northern Hemisphere.

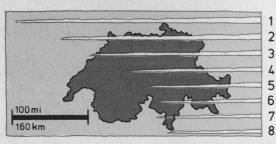

100 mi
160 km

1
2
3
4
5
6
7
8

Mountain ranges

1	Cordillera de Los Andes	S America	4500mi
2	Rocky Mountains	N America	3750mi
3	Himalayas–Karakoram–Hindu Kush	Asia	2400mi
4	Great Dividing Range	Oceania	2250mi
5	Trans-Antarctic Mountains	Antarctica	2200mi
6	Brazilian East-Coast Range	S America	1900mi
7	Sumatran-Javan Range	Asia	1800mi
8	Aleutian Range	N America	1650mi

Rivers

1	Nile	Africa	4132mi
2	Amazon	S America	3900mi
3	Mississippi–Missouri–Red Rock	N America	3860mi
4	Ob-Irtysh	Asia	3461mi
5	Yangtze	Asia	3430mi
6	Hwang Ho	Asia	2903mi
7	Congo (Zaire)	Africa	2900mi
8	Amur	Asia	2802mi
9	Lena	Asia	2653mi
10	Mackenzie	N America	2635mi
11	Mekong	Asia	2600mi
12	Niger	Africa	2590mi
13	Yenisey	Asia	2566mi
14	Paraná	S America	2450mi
15	Plata-Paraguay	S America	2300mi
16	Volga	Europe	2293mi
17	Madeira	S America	2060mi
18	Indus	Asia	1980mi

Glaciers

1	Lambert-Fisher	Antarctica	320mi
2	Novaya Zemlya	USSR	260mi
3	Arctic Institute	Antarctica	225mi
4	Nimrod-Lennox-King	Antarctica	180mi
5	Denman	Antarctica	150mi
6	Beardmore	Antarctica	140mi
7	Recovery	Antarctica	140mi
8	Petermanns	Greenland	124mi

"In the space of one hundred and seventy-six years the Lower Mississippi has shortened itself two hundred and forty-two miles. That is an average of a trifle over one mile and a third per year. Therefore . . . just over a million years ago next November, the Lower Mississippi River was upward of one million three hundred thousand miles long, and stuck out over the Gulf of Mexico like a fishing-rod. And by the same token . . . seven hundred and forty-two years from now the Lower Mississippi will be only a mile and three-quarters long, and Cairo and New Orleans will have joined their streets together. . . . There is something fascinating about science. One gets

Although Europe is the second smallest continent in terms of area, it has the second longest coastline (37,887 miles).

Global girdle *right*
The world's longest natural feature is the submarine mountain range that runs down the Atlantic, through the Antarctic. and on to end in the NE Pacific. If straightened, its 40,000mi would go more than 1½ times around the Equator.

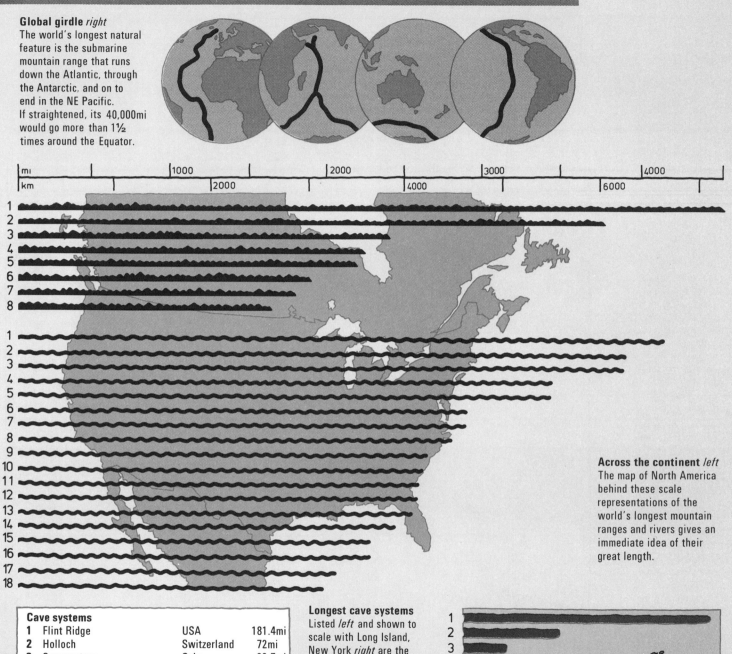

Across the continent *left*
The map of North America behind these scale representations of the world's longest mountain ranges and rivers gives an immediate idea of their great length.

Cave systems		
1 Flint Ridge	USA	181.4mi
2 Holloch	Switzerland	72mi
3 Cuyaguatega	Cuba	32.7mi
4 Werfen	Austria	26.1mi
5 Peschtschera	USSR	22.7mi
6 Palomera	Spain	22.5mi
7 Ogof Ffynnon	UK	20.3mi
8 Postojnska	Yugoslavia	17mi

Longest cave systems
Listed *left* and shown to scale with Long Island, New York *right* are the longest cave systems of eight countries. Linking the Mammoth Cave System to the Flint Ridge System, Kentucky in 1972 made this the world's longest.

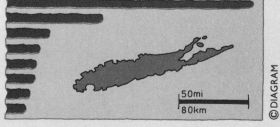

©DIAGRAM

such wholesome returns of conjecture out of such a trifling investment of fact." (Extract from *Life on the Mississippi* by Mark Twain.)

The longest cave system in the world, Flint Ridge, USA, is some 80 miles longer than the tunnels of the world's longest subway system, in London, England (101 miles).

BUILDINGS AND MONUMENTS

The United States boasts the world's tallest apartment block, tallest hotel, tallest office block, and several of the world's tallest monuments. It is, however, to Europe that we must turn for the tallest of all manmade structures—the Warsaw Radio mast, some 57ft taller than the KTHI-TV mast in Fargo, N Dakota.

Tall stories *right*
A Ulm cathedral, W Germany, the world's tallest cathedral, 528ft.
B World's largest cooling tower, at Uentrop, W Germany, 590ft.
C Lake Point Towers, Chicago, tallest apartment block, 645ft.
D Peachtree Center Plaza, Atlanta, tallest hotel, 723ft.
E Chrysler Building, New York, 1046ft.
F Tallest chimney, International Nickel Co, Sudbury, Ontario, 1245ft.
G Empire State Building, New York, 1250ft without mast, 1472ft with mast.
H World Trade Center, New York, 1353ft.
I Sears Tower, Chicago, since 1973 the tallest office building in the world, 1454ft without mast, 1559ft with mast.
J CN Tower, Toronto, the tallest self-supporting tower, 1815ft.
K Warsaw Radio mast, near Płock, Poland, the tallest structure in the world, 2120ft 8in.

When building the Tower of Babel, the descendants of Noah sought to build a tower "whose top may reach unto heaven." To reach only as high as Mount Everest (29,002ft), a tower would have to be 20 times the height of Sears Tower (1454ft without mast).

All to scale *left*
Shown here to a common scale are the various man-made structures included on these two pages. The Warsaw Radio mast (**a**) and the Eiffel Tower (**b**) have been drawn outside the boxes to give a view of their full heights.

Mighty monuments *left*
1 Cleopatra's Needle, now in London, 68ft.
2 Arch of Septimus Severus, Rome, 68ft.
3 Trajan's Column, Rome, 115ft.
4 Tallest totem pole, at Alert Bay, Canada, 173ft.
5 Motherland sculpture, Volgograd, USSR, world's tallest free-standing statue, 270ft.
6 Statue of Liberty, New York, height from sandals to top of torch 151ft, height with pedestal 305ft.
7 Great Pyramid of Cheops, Giza, Egypt, original height 480ft 11in.
8 Washington Memorial, Washington DC, 555ft.
9 San Jacinto Column, Texas, world's tallest monumental column, 570ft.
10 Gateway to the West Arch, St Louis, Missouri, 630ft.
11 Eiffel Tower, Paris, original height 985ft 11in, since addition of TV mast 1052ft 4in.

©DIAGRAM

The Statue of Liberty, New York, is approximately 20 times life size; the Motherland sculpture, Volgograd, is more than 30 times life size.

The top platform of the Eiffel Tower is the same height as the 73rd floor of the Empire State Building; the Empire State's own observatories are on the 86th and 102nd floors.

DRILLING AND MINING

Man's search for clean water, coal, precious metals and stones, gas and oil has led him to delve deep into the Earth's crust. Here we compare some of the depths to which he has gone—down some 12,600ft (or over 2 miles) in the deepest mine, and down some 31,911ft (or just over 6 miles) with the deepest drilling.

A) Deep mining		
1 Kolar (1919)	India	5419ft
2 Witwatersrand (1931)	S Africa	7640ft
3 Nova Lima (1933)	Brazil	8051ft
4 Witwatersrand (1933)	S Africa	8198ft
5 Witwatersrand (1934)	S Africa	8400ft
6 Johannesburg (1938)	S Africa	8527ft
7 Kolar (1939)	India	8604ft
8 Johannesburg (1949)	S Africa	9071ft
9 Boksburg (1953)	S Africa	9288ft
10 Boksburg (1958)	S Africa	11,000ft
11 Boksburg (1959)	S Africa	11,246ft
12 Western Deep (1975)	S Africa	12,600ft
B) Deep drilling		
1 Szechwan (c150BC)	China	2000ft
2 Olinda (1927)	California	8046ft
3 Vera Cruz (1931)	Mexico	10,585ft
4 Belridge (1934)	California	11,377ft
5 Pecos County (1944)	Texas	15,279ft
6 Caddo County (1947)	Oklahoma	17,823ft
7 Sublette County (1949)	Wyoming	20,521ft
8 Bakersfield (1953)	California	21,482ft
9 Plaquemines (1956)	Louisiana	22,570ft
10 Pecos County (1958)	Texas	25,340ft
11 St Bernard Parish (1970)	Louisiana	25,600ft
12 Pecos County (1972)	Texas	28,500ft
13 Beckham County (1972)	Oklahoma	30,050ft
14 Washita County (1974)	Oklahoma	31,441ft
15 Kola peninsula (1979)	USSR	31,911ft

Highest and lowest *above*
The highest man-made structure is the Warsaw Radio mast (**a**), at 2120ft 8in; the tallest office building is the Sears Tower, Chicago (**b**), at 1454ft without mast. The lowest at which men work is 12,600ft, in the world's deepest mine (**c**).

Deep mining and drilling
Listed in the table *above* and drawn to scale *far right* are some of the depths that man has reached by mining (**A**) and drilling(**B**). The date of the achievement is given in brackets after the name of the mine or drilling site.

Digging deep *right*
This diagram shows to scale some of man's excursions into the Earth's crust. The greatest depth at which a man has worked is 12,600ft, but drilling machinery has been operated at more than 2½ times this depth, at 31,911ft.

The world's deepest mine is nearly three times as deep as the world's deepest known cave.

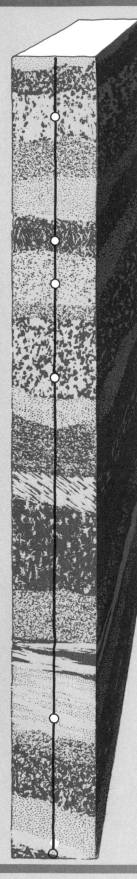

−2540ft the world's deepest open mine

−7320ft the deepest water well in the world, in Montana, USA

−9029ft the world's deepest steam well, in California, USA

−12,600ft the deepest mine in the world, Western Deep, S Africa

−26,192ft the greatest depth reached during the Moho project (5709ft into the seabed)

31,911ft the world's deepest drilling (July 1979), on the Kola peninsular, USSR

If made from the top of Mount Everest, the world's deepest drilling would extend some 2909ft below sea level.

A Deep mining *left*
Shown here to scale are some of man's deepest mines, all of which are gold mines. The deepest (12,600ft) is nearly six times the height of the world's tallest structure— the Warsaw Radio mast.

B Deep drilling *left*
Shown to a smaller scale are various depths achieved by drilling into the ground. The deepest drilling (31,911ft) exceeds the depth of the deepest mine by 19,311ft – over 13 times the height of Sears Tower (1454ft without mast) or 9 times the height of the Warsaw Radio mast (2120ft 8in).

Structures to scale *left*
Included on our mining and drilling diagrams to illustrate the scales are:
a Warsaw Radio mast (2120ft 8in)
b Sears Tower (1454ft without mast, 1559ft with mast)
c Empire State Building (1250ft without mast, 1472ft with mast)

41

©DIAGRAM

BRIDGES, TUNNELS AND CANALS

A suspension bridge with a single span some seven-eighths of a mile wide, a ship canal deep enough to take sea-going vessels for a distance in excess of 140 miles, and a 17-mile-long tunnel in the London underground railway system are among the greatest feats of engineering devised in the face of daunting obstacles to efficient transportation.

Lengths of bridge spans
right Listed are the five longest examples of three types of bridge span—suspension, cantilever, and steel arch. The longest span for a suspension bridge is over 2½ times that for any other type of bridge.

Bridge spans		
Suspension		
1 Humber Estuary*	Humberside, UK	4626ft
2 Verrazano Narrows	New York, USA	4260ft
3 Golden Gate	San Francisco, USA	4200ft
4 Mackinac Straits	Michigan, USA	3800ft
5 Atatürk	Istanbul, Turkey	3524ft
Cantilever		
1 Quebec	Quebec, Canada	1800ft
2 Firth of Forth	Midlothian/Fife, UK	1710ft
3 Delaware River	Pennsylvania, USA	1644ft
4 Greater New Orleans	Louisiana, USA	1575ft
5 Howrah	Calcutta, India	1500ft
Steel arch		
1 New River Gorge	W Virginia, USA	1700ft
2 Bayonne	New Jersey/ New York	1652ft
3 Sydney Harbour	Sydney, Australia	1650ft
4 Fremont	Oregon, USA	1255ft
5 Port Mann	Vancouver, Canada	1200ft
*Due for completion 1980		

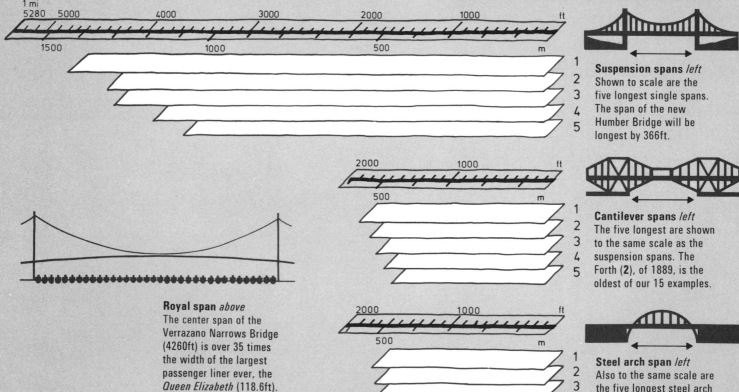

Suspension spans *left*
Shown to scale are the five longest single spans. The span of the new Humber Bridge will be longest by 366ft.

Royal span *above*
The center span of the Verrazano Narrows Bridge (4260ft) is over 35 times the width of the largest passenger liner ever, the *Queen Elizabeth* (118.6ft).

Cantilever spans *left*
The five longest are shown to the same scale as the suspension spans. The Forth (**2**), of 1889, is the oldest of our 15 examples.

Steel arch span *left*
Also to the same scale are the five longest steel arch spans. The theoretical limit for a span of this type is about 3280ft.

A bridge 21 miles long would be needed to traverse the English Channel between Dover and Calais; the world's longest bridging to date is the 23.8–mile–long Second Lake Pontchartrain Causeway in Louisiana.

The world's longest vehicular tunnel, London Transport's Northern Line, would stretch under the Mediterranean Sea from the Rock of Gibraltar to a point in Morocco some 3 miles from the coast.

Tunnels		
1 Northern Line (subway)	London, UK	17.3mi
2 Simplon I, II (rail)	Switzerland/Italy	12.3mi
3 Shin Kanmon (rail)	Japan	11.6mi
4 Gt Apennine (rail)	Italy	11.5mi
5 St Gotthard (road)*	Switzerland	10.1mi
6 Rokko (rail)	Japan	10.0mi
7 Henderson (rail)	USA	9.8mi
8 Lötschberg (rail)	Switzerland	9.0mi
Canals		
1 White Sea—Baltic	USSR	141mi
2 Suez	Egypt	100.6mi
3 Volga—Don	USSR	62.2mi
4 North Sea	Germany/Denmark	60.9mi
5 Houston	USA	56.7mi
6 Panama	Panama	50.7mi
7 Manchester Ship	UK	39.7mi
8 Welland	Canada	28.0mi

*Due for completion 1980

Lengths of tunnels
The table *left* and the diagram *below* show the eight longest vehicle-carrying tunnels in the world. The longest non-vehicular tunnel, the Delaware Aqueduct in New York State, is 105 miles in length.

Largest diameter tunnel
right The tunnel through Yerba Buena Island in San Francisco is 76ft wide and 58ft tall (or over 11 typical cars wide and 12 typical cars high).

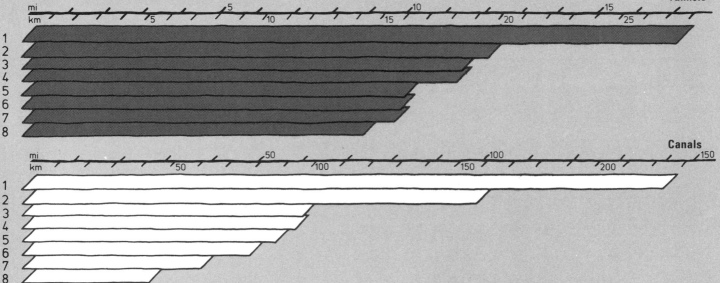

Lengths of ship canals
Listed in the table *top* and shown to scale in the diagram *above* are the world's eight longest deep-draft ship canals (minimum depth 16.4ft). Suez, the oldest major ship canal (1869), is still the second longest.

Longest canal system
right The longest canal system in the world is the Volga-Baltic Canal, from Astrakhan on the Caspian Sea (**A**) to Leningrad (**B**)— a distance of 1850mi, equal to the road distance from Leningrad to the edge of Paris (**C**).

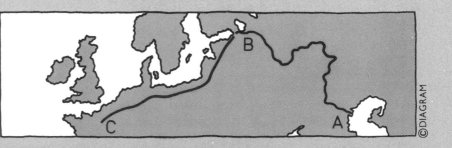

©DIAGRAM

If the canals of the Netherlands (2182 miles in all) had been dug end to end along a Great Circle from New Amsterdam (or, as it is now known, New York), they would extend 97 miles beyond Mexico City.

SHIPS AND BOATS

Throughout history man has designed ships to transport passengers and cargo and to serve as weapons of war. Shown below to a common scale are civil and military ships from different periods. Vessels of the new generation of supertankers, represented here by the *Bellamya*, are longer than the world's longest aircraft carrier.

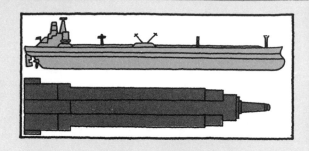

Lengths of merchant and passenger ships *below*
A *Bellamya*, French oil tanker, 1312ft. (The largest ship afloat is another tanker, the *Pierre Guillaumat*, 1359ft.)
B 19th-century whaling ship, 107ft.

C RMS *Queen Elizabeth*, launched 1938, the largest liner ever, 1031ft. (Today's largest, the QE2, is 963ft.)
D SS *Great Eastern*, 1858, passenger ship, largest vessel until 1899, 692ft.
E *Natchez*, 1869, Mississippi packet, 307ft.

F Egyptian merchant ship, c.1500 BC, 90ft.
G Soling class yacht, 26ft 9in.

Long and tall
above The supertanker *Bellamya* (1312ft) is 62ft longer than the Empire State Building is tall (1250ft without mast).

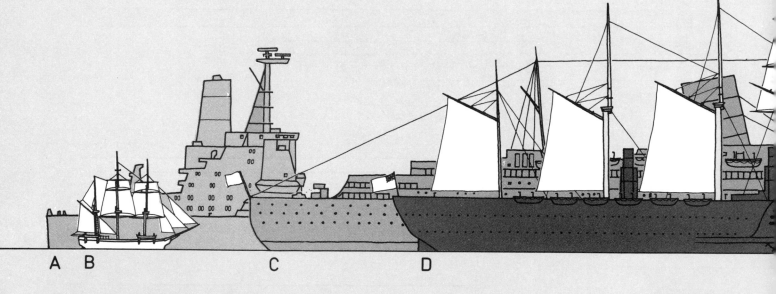

A B   C   D

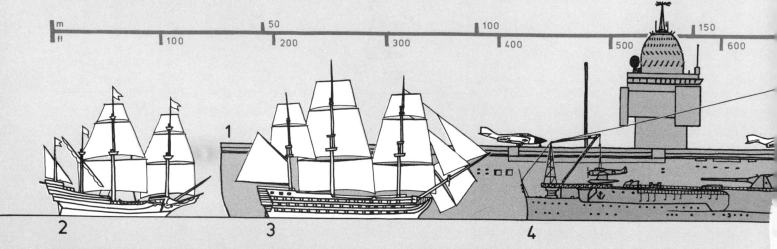

m		50		100		150
ft	100	200	300	400	500	600

2   3   4

Noah was a remark-
able man according to
the Bible, for with his
three sons he built the
450ft-long ark—twice
as long as Nelson's
Victory.

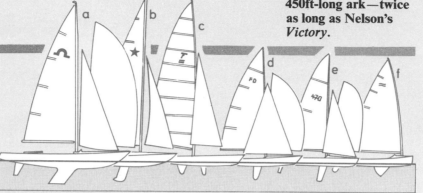

Lengths of warships
See illustrations *below*.
1 USS *Enterprise*, 1960,
nuclear-powered aircraft
carrier, the longest
warship ever, 1123ft.
2 English warship of the
17th century, 142ft.

3 HMS *Victory*, Nelson's
flagship at Trafalgar in
1805, 226ft.
4 *Yamato*, Japanese WW2
battleship. With its sister
Musashi, the largest but
not the longest battleship
ever, 863ft.

5 USS *Monitor*, 1862,
ironclad warship of the
American Civil War, 172ft.
6 Roman galley, 235ft.

Olympic yachts
above Lengths for yachts
in the six Olympic classes
for 1980 are :
a Soling 26ft 9in
b Star 22ft 8in
c Tornado 20ft
d Flying dutchman 19ft 10in
e "470" 15ft 4¾in
f Finn 14ft 9in

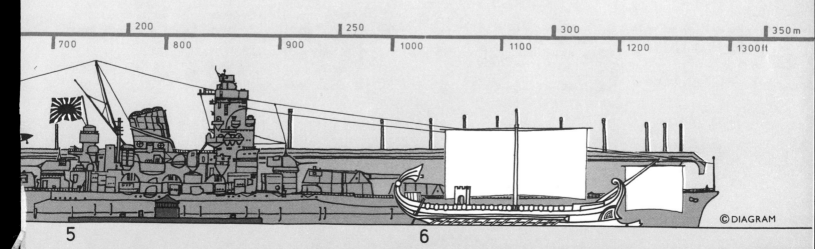

E   F   G

200	250	300	350m

700	800	900	1000	1100	1200	1300ft

5   6

©DIAGRAM

LAND VEHICLES

Drawn to scale on these two pages is a selection of the world's largest land vehicles, civil and military, mobile and partly mobile. The Rosenkranz K10001 crane is about two thirds the height of the Eiffel Tower. The world's longest car, a special Cadillac, is 10ft 2in longer than a regular Cadillac Brougham.

Large land vehicles
A Specially built (1976) Cadillac, the longest car ever made, 29ft 6in.
B Transporter used to carry Saturn V rockets at the John F. Kennedy Space Center, Florida. The most massive vehicle ever built, it measures 131ft 4in by 114ft. (See overleaf for a comparison of the 363ft 8in-tall Saturn V with selected aircraft.)
C The longest known freight train ran in 1968 from Iaeger, W Virginia, to Portsmouth, Ohio. Made up of 500 coal cars and six engines, three pulling and three pushing, it was about 4 miles long.
D Paris gun from World War 1, used to fire 275lb shells on Paris from behind enemy lines about 70 miles away. Height of gun carriage 25ft. Length of gun barrel 128ft.

E Chieftain tank, the heaviest and one of the longest tanks currently in service. Length of hull 24ft 8in.
F The world's tallest mobile crane, the Rosenkranz K10001, has a combined boom and jib height of 663ft. It can lift 29½ tons to a height of 525ft.
G England's Jodrell Bank telescope has a fully steerable dish, 250ft in diameter. (The telescope with the largest trainable dish, 328ft in diameter, is in the Effelsberger Valley, W Germany.)

G

F

All to scale *above*
A smaller scale is used here to allow us to compare in their entirety the Saturn V rocket, on its transporter, the Jodrell Bank telescope and the world's largest mobile crane.

AIRCRAFT

The combined lengths of Concorde and
the Boeing B-52H bomber fall just short of
the length of the Saturn V space rocket.
Yet Saturn V itself would fit more than
twice into the length, when inflated, of the
airship *Hindenburg*, which wafted passengers
in great comfort across the Atlantic in the
years before World War 2.

A

C

B

Up, up and away !
All these aircraft are drawn
to the same scale.
A Saturn V rocket, used for
the Apollo and Skylab
missions, 363ft 8in tall
including module.
B Boeing B-52H, the world's
heaviest bomber, wing span
185ft, length 157ft 7in.
C Airship *Hindenburg*,
809ft 5in long (only half
its length is shown here).
D Montgolfier balloon of
1783, diameter of envelope
38ft, height 49ft.
E Boeing 747 Jumbo Jet,
biggest capacity jet
airliner, wing span 195ft
8in, length 231ft 4in.
F MiG-25 fighter, wing span
45ft 9in, length 73ft 2in.
G Concorde, supersonic
airliner, wing span 83ft 10in,
length 203ft 9in.

D

Grounded Jumbo *below*
A Jumbo Jet (231ft 4in
long, 195ft 8in wing span)
is shown on a soccer pitch
(330ft long, 240ft wide).

If the airship *Hindenburg* had been stood
on its end next to the Empire State Building,
its nose would have reached almost to the
windows on the 62nd floor.

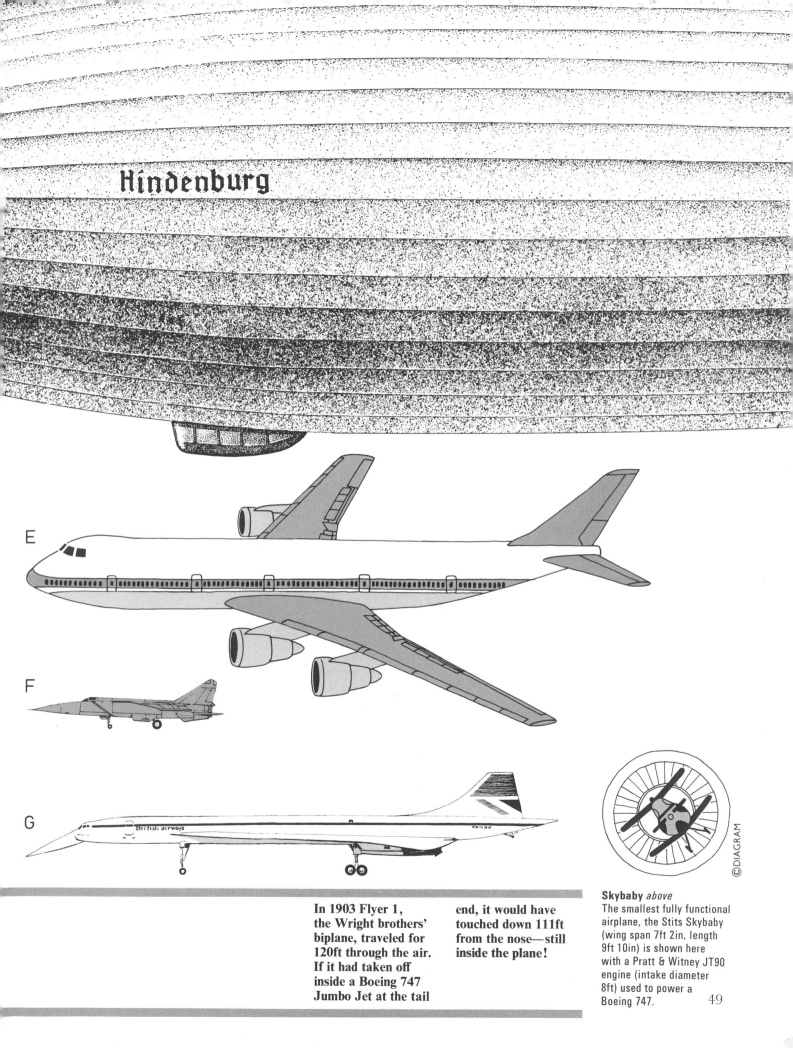

Hindenburg

E

F

G

In 1903 Flyer 1,
the Wright brothers'
biplane, traveled for
120ft through the air.
If it had taken off
inside a Boeing 747
Jumbo Jet at the tail

end, it would have
touched down 111ft
from the nose—still
inside the plane!

Skybaby *above*
The smallest fully functional
airplane, the Stits Skybaby
(wing span 7ft 2in, length
9ft 10in) is shown here
with a Pratt & Witney JT90
engine (intake diameter
8ft) used to power a
Boeing 747.

49

THE SOLAR SYSTEM

The Sun is much larger than any of the planets in orbit around it. Even Jupiter, the largest planet, has an equatorial diameter only about one tenth that of the Sun. Earth, which ranks fifth among the planets in terms of size, has an equatorial diameter less than one tenth that of Jupiter and one hundredth that of the Sun.

 2   3   4   5

Sun	865,500mi	1,392,900km	
Mercury	3032mi	4880km	0
Pluto	3700mi	6000km	0
Mars	4217mi	6787km	2
Venus	7521mi	12,104km	0
Earth	7926mi	12,756km	1
Neptune	30,800mi	49,500km	2
Uranus	32,200mi	51,800km	5
Saturn	74,600mi	120,000km	10
Jupiter	88,700mi	142,800km	13

Solar sisters *above*
The scale drawings across the top of these pages and the tabular material on equatorial diameters allow us to compare the sizes of the Sun and its planets. We have included both imperial and metric measurements on these pages because scientific data is now more usually expressed in metric units. The most up-to-date information available has been used, but figures are subject to frequent revision. Also shown here are the known number of satellites for each planet.

Imperfect world *left*
Like other planets the Earth is not a perfect sphere, being flattened slightly at the top and bottom. Measured at the Equator, its diameter is 7926mi (12,755km); at the Poles the diameter is only 7900mi (12,713km).

On a human scale *above*
If we suppose that the Sun's diameter is equal to the height of an average man, then Jupiter, the largest planet, would be slightly smaller than the man's head, while Earth would be slightly bigger than the iris of his eye.

Size of asteroids *right*
Asteroids are rocky bodies with diameters ranging from a few feet to several hundred miles. Examples are shown in scale with California.
a Ceres 429mi (690km)
b Vesta 244mi (393km)
c Fortuna 100mi (161km)
d Eros 11mi (18km)

Measuring only 6 miles by 7½ miles by 10 miles, Deimos, the smaller of Mars's tiny rocky moons, would fit onto the John F. Kennedy Space Center, Cape Canaveral, Florida.

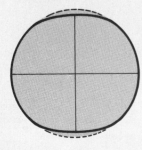

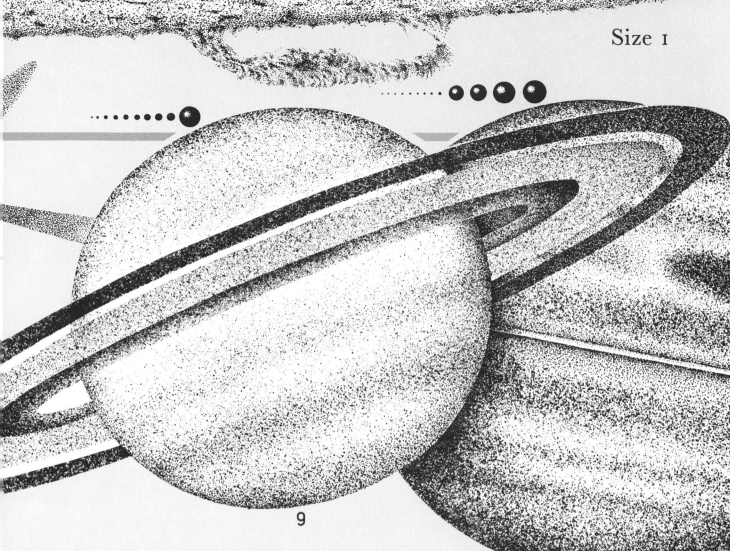

9

10

Earth and Moon *below*
For its size, Earth (**1**) has a larger satellite, or ''moon'' (**2**), than any other planet in our solar system. Earth's equatorial diameter is 7926mi (12,755km), while the Moon's diameter is just over one quarter of that, 2158mi (3473km).

Large moons *below*
Though much smaller than their own planets, these are bigger than Mercury:
a Titan, a moon of Saturn, 3600mi (5800km)
b Ganymede, a moon of Jupiter, 3275mi (5270km)
c Triton, a moon of Neptune, roughly 3000mi (4800km)

Mighty meteorite *below*
Meteorites—pieces of rock drawn into Earth's atmosphere from space— usually burn up before landing. The largest known to have landed, drawn here with a man of average size, is 9ft (2.74m) long by 8ft (2.44m) wide.

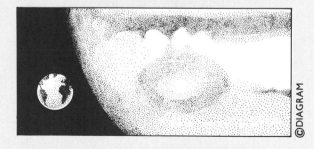

©DIAGRAM

Great Red Spot *above*
Often visible in Jupiter's atmosphere is the Great Red Spot, thought to be an anticyclone with a difference, for this one is some 8000mi (13,000km) wide and 25,000mi (40,000km) long—more than three times the diameter of Earth.

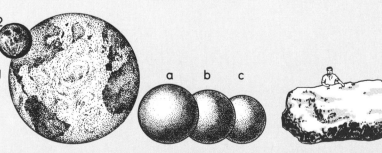

At a steady jogger's pace of 6mph it would take 173 days to go around the equatorial circumference of Earth, and more than 5 years (some 1935 days) to go around the circumference of the largest planet, Jupiter.

SUN AND STARS

The Sun is over 100 times bigger than Earth, and in Earth terms solar features like flares and sunspots are of truly gigantic proportions. In galactic terms, however, our Sun is only one of perhaps 100,000 million stars, which range in size from half as big as our Moon to over 10,000 times larger than Earth.

Relative sizes *right*
If the diameter of the largest known star, IRS5, were the height of Mount Everest, our Sun's diameter would be the height of an 18-month-old child, and Sirius B, one of the smallest known stars, would be only 1in in diameter.

1

Solar flares *above*
Violent eruptions of incandescent gases, solar flares are an impressive feature of the Sun's atmosphere. Huge arcs of gas disintegrate and stream out into space (**1**), to distances greater than that from Earth to Moon.

Sunspots *left*
Groups of sunspots are a common feature of the Sun's surface, sometimes visible to the naked eye. A single spot may measure as much as 8 times the Earth's diameter. Groups are typically 100,000mi (160,000km) across.

Star story *right*
At its birth the Sun was a small red star, formed by the contraction of interstellar material. It is now in the second stage of its development, a fairly small brilliant star, represented here by the small sphere in the center. Over a period of perhaps 10,000 million years it is expected to expand, as shown here, until its diameter is 50 times its present size. It will then become unstable, eject matter and collapse into a small dense white dwarf, smaller than it is now.

Little star *left*
The smallest known star is the white dwarf LP 327–186. As shown here, its estimated diameter, approximately 1000mi (1600km), is only one-half that of our Moon.

Over 63 million stars the size of the largest known star (IRS5, estimated diameter 9,200,000,000 miles) would need to be placed side by side to cover a distance equal to the estimated diameter of our galaxy (100,000 light years).

Seeing stars !
Shown here to scale and included in the data table *below* are the Sun and some much larger stars. The figure in brackets after the name of each star is the number of times by which the estimated diameter of that star is bigger than that of the Sun. This is followed by the approximate diameter of each star, given in both miles and kilometers.

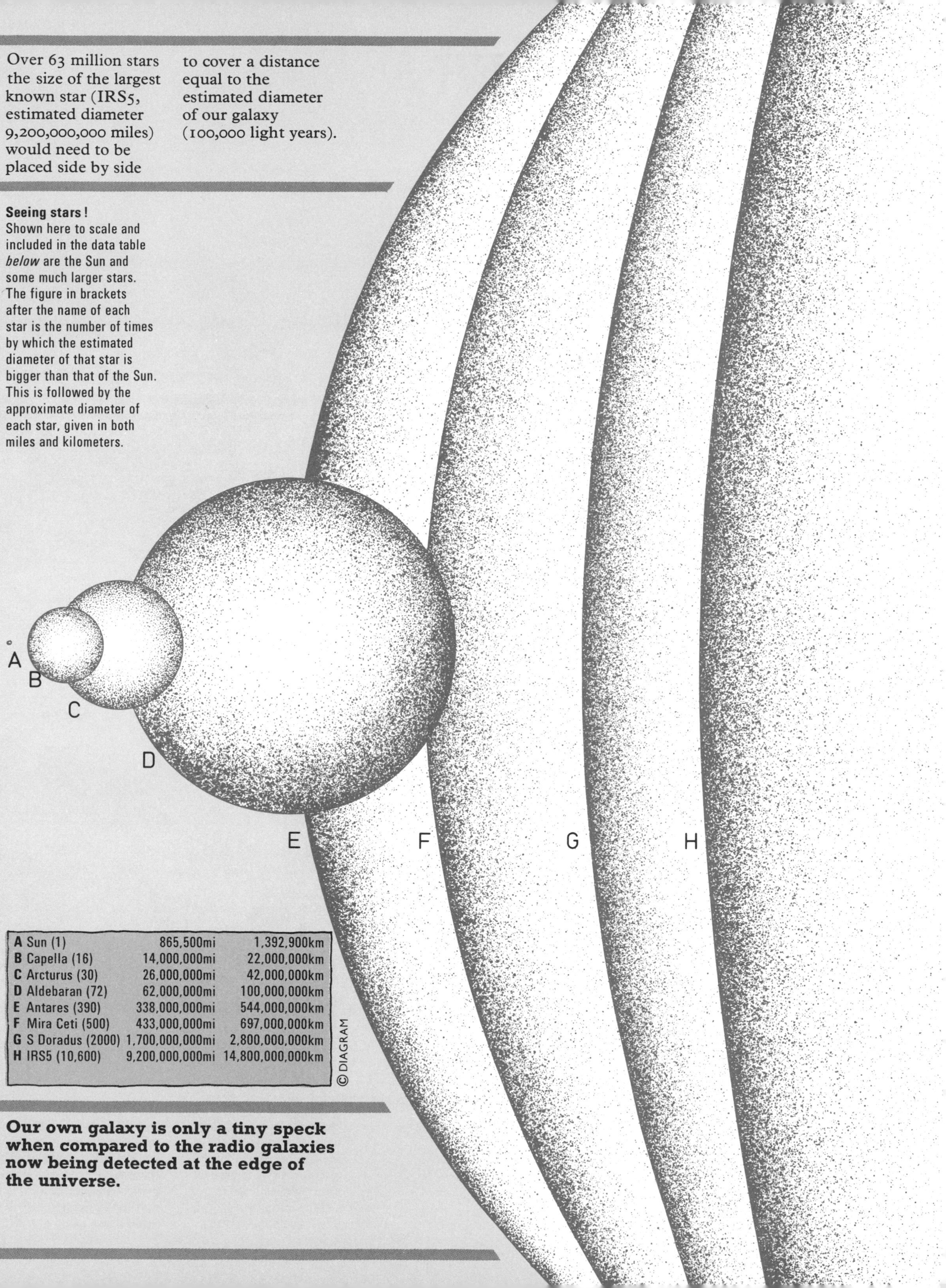

A Sun (1)	865,500mi	1,392,900km
B Capella (16)	14,000,000mi	22,000,000km
C Arcturus (30)	26,000,000mi	42,000,000km
D Aldebaran (72)	62,000,000mi	100,000,000km
E Antares (390)	338,000,000mi	544,000,000km
F Mira Ceti (500)	433,000,000mi	697,000,000km
G S Doradus (2000)	1,700,000,000mi	2,800,000,000km
H IRS5 (10,600)	9,200,000,000mi	14,800,000,000km

©DIAGRAM

Our own galaxy is only a tiny speck when compared to the radio galaxies now being detected at the edge of the universe.

CHAPTER 3

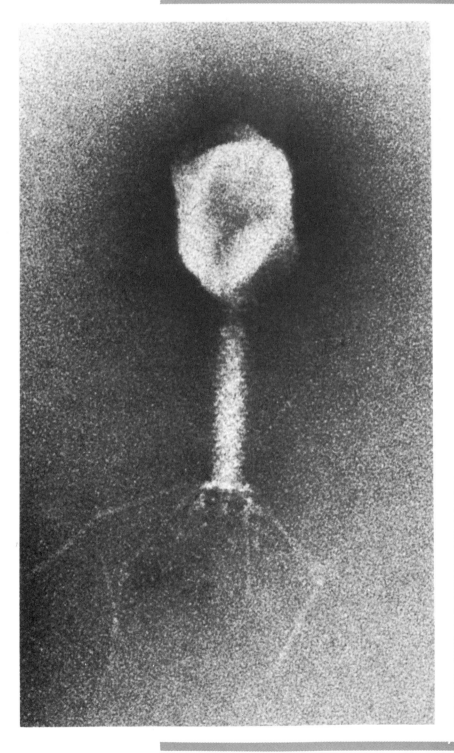

A T4 virus, which attacks the bacteria of the colon, is shown more than 400,000 times actual size in this enlarged negative-contrast electron micrograph (Basel University, Switzerland).

Post card showing a huge Californian water melon— evidence of an unofficial record-breaker, or a clever photomontage (Fred C. Moran; Smith International News, Los Angeles).

An 18th-century English engraving showing whale-men at work in Greenland. The whale's great size and strength are insufficient protection against the weapons and greed of man.

SIZE: LIVING WORLD

A publicity photograph for
Giant Machnow, an
Irishman who appeared at
the London Hippodrome in
1905. He was billed as
being some 9ft 4in tall
(Mansell Collection).

MICROSCOPIC LIFE

Microscopes and electron microscopes allow us to see in detail a host of living organisms that are invisible to the naked eye. On these two pages we use different degrees of magnification to compare some of these organisms, both plant and animal, unicellular and multicellular, simple and exceedingly complex.

Small-scale creatures
below Progressively greater magnifications are here used to show smaller and smaller life-forms.
Magnification A (x100)
1 Euglena, which has plant and animal characteristics.
2 Amoeba, a minute animal.
3 Chlamydomonas, a plant.

Magnification B (x1000)
3 Chlamydomonas, with its whiplike flagella, can now be seen quite clearly.
4 A bacterium, however, is still only a tiny speck when magnified 1000 times.

Not the bee's knees!
left Here we compare the relative sizes of a queen honeybee and a parasite that lives upon it.
a A queen honeybee (1.5cm long) is shown here actual size. Drones are similar in size to the queen, but workers are smaller (1.2cm).
b One of the honeybee's legs is here magnified 10 times; the parasite (drawn in color) is also visible at this magnification.
c A magnification of 100 times actual size gives a detailed picture of the honeybee's unwelcome guest: *Braula caeca* (0.8mm long).

a

1cm

b

1mm

c

·1mm

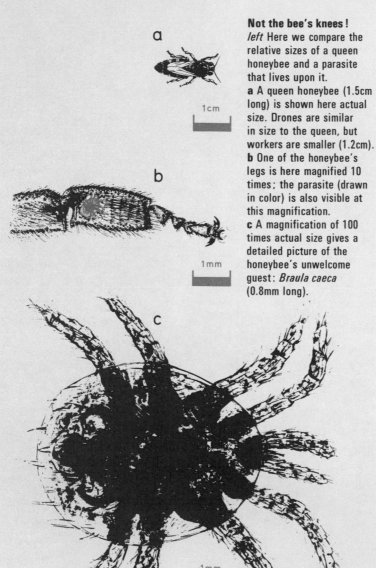

1

2

3

·1mm

A

Microscopic monsters
right A magnification of 100 times actual size (similar to that used for the honeybee parasite *left*) shows huge variety among microscopic creatures. Here we show a small selection.
d *Phthirus pubis* (1–1.5mm long), crab louse, a blood-sucking parasite of humans.
e *Tetranychus urticae* (0.5mm), a reddish mite that sucks plant juices.
f *Folsomia fimetaria* (1mm), a springtail that lives in soil and attacks roots.
g *Knemidocoptes mutans* (0.25mm male; 0.45mm female), mange mite, gnaws the skin of many animals.

d

Chondromyces bacteria climb one on top of another to form towers several millimeters high. On a human scale, these towers would be over a mile high, or some four times the height of the World Trade Center in New York.

Magnification C (x10,000)
At this magnification we can clearly distinguish examples of the three basic types of bacterium.
4 Cocci are spherical, and may be in groups or chains.
5 Spirilla are spiral or comma-shaped.
6 Bacilli are rod-shaped.

Magnification D (x100,000)
6 Bacillus, with details of its structure now clearly visible.
7 Virus, shown attacking the bacillus; viruses can reproduce only inside other living bodies.
Magnification E (x1,000,000)
7 Part of the same virus.

Quick scale-guide *left*
Here we use well-known creatures to show the relative sizes of some of the life-forms *below*. If an amoeba (**2**) were as big as an elephant, chlamydomonas (**3**) would be the size of a cat, and a bacterium (**4**) the size of a flea.

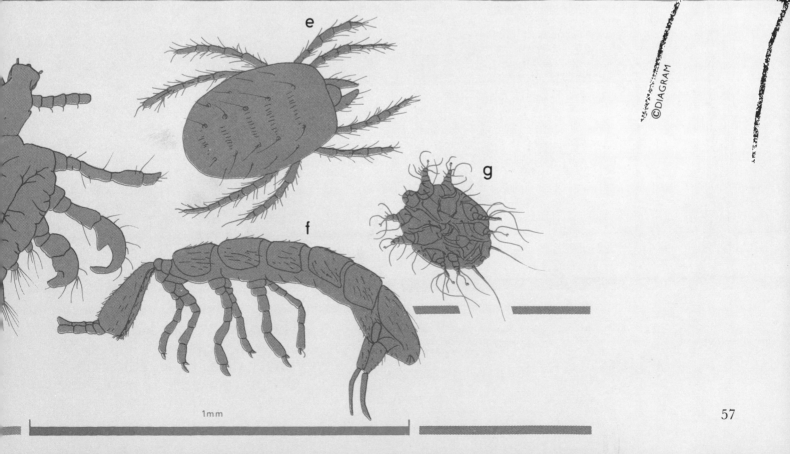

3

4

4

5

6

6

7

7

·01mm
B

·001mm
C

·0001mm
D

·00001mm
E

©DIAGRAM

e

f

g

1mm

PLANTS

On the previous two pages we compared some of the very smallest living things. Here we turn to some of the largest, all members of the plant kingdom. The Californian redwood is the tallest species of plant alive today. Also illustrated for comparison are some unusually large examples of generally much smaller plants.

Tallest tree *left*
Trees are the tallest of all living things, and the Californian redwood (1) is the tallest of all modern trees. Its record height of 366ft exceeds the length of a soccer pitch by 36ft, and is over 63 times the height of an average man.

Tallest plants *below*
Some record plants might not seem particularly tall when compared with the enormous Californian redwood, but each of the examples included here is a giant of its own kind.
1 Californian redwood 366ft
2 Bamboo 121ft
3 Tree fern *Alsophila excelsa* 60ft
4 Saguaro cactus 52ft
5 Orchid *Grammatophyllum speciosum* 25ft
6 Callie grass 18ft

Floral phenomena *right*
How about considering one or two of these splendid specimens for an impressive floral display? They are drawn here to scale along with a gardener, a man of average height (5ft 9in).
A The world's tallest orchid, *Grammatophyllum speciosum* from Malaysia, is sometimes 25ft tall.
B The tallest sunflower measured in the UK was 23ft 6½in high.
C The US record height for a hollyhock is an imposing 18ft 9½in.
D The tallest recorded dahlia measured 9ft 10¾in and was grown in Australia.
E The British record for a lupin is 6ft 0½in.

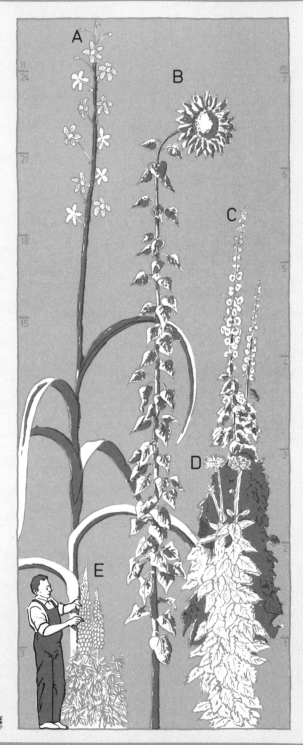

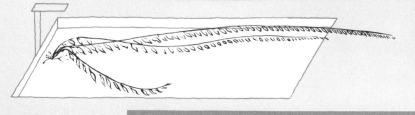

The longest recorded strand of seaweed measured 196ft, 31ft longer than the length of an Olympic swimming pool.

To quote a poem by David Everett (1769–1813), "tall oaks from little acorns grow," but to be more precise, 120ft oak trees grow from ¾in long acorns.

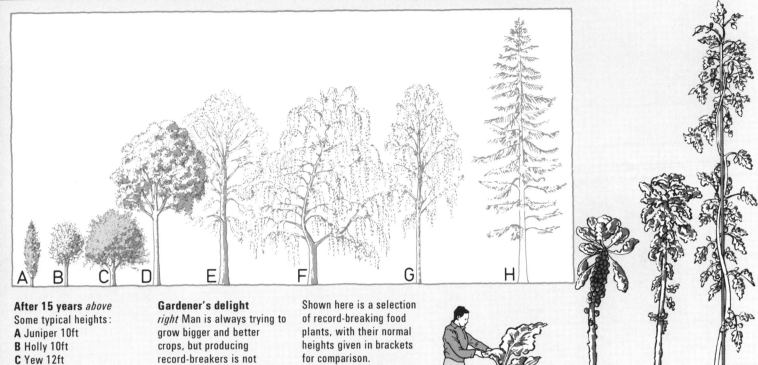

After 15 years *above*
Some typical heights:
A Juniper 10ft
B Holly 10ft
C Yew 12ft
D Oak 25ft
E Poplar 30ft
F Weeping willow 30ft
G Silver birch 30ft
H Douglas fir 40ft

Gardener's delight
right Man is always trying to grow bigger and better crops, but producing record-breakers is not without its problems. Imagine having to climb a ladder to pick a Brussels sprout from a plant that is over 10ft tall!

Shown here is a selection of record-breaking food plants, with their normal heights given in brackets for comparison.
1 Rhubarb 5ft 1in (2–3ft)
2 Brussels sprout 10ft 8in (3ft)
3 Kale 12ft (1–3ft)
4 Tomato 20ft (3–4ft)

Towering timber *left*
All these record-breaking trees come from the USA.
A Californian redwood 366ft
B Douglas fir 302ft
C Fir noble 278ft
D Giant sequoia 272ft
E Ponderosa pine 223ft
F Cedar 219ft
G Sitka spruce 216ft
H Western larch 177ft
I Hemlock 163ft
J Beech 161ft
K Black cottonwood 147ft

This letter o could contain over 20 blooms of the artillery plant *Pilea microphylla* from India. The smallest blooms known, their diameter is only 1/72 of an inch.

The largest bloom is that of the stinking corpse lily *Rafflesia arnoldii*. Its diameter of up to 3ft is equal to the width of 4½ pages of this book.

MARINE CREATURES

In medieval times, sailors were terrified by the humble barnacle, thought to be an enormous creature that ate the hulls of ships. Today, cinema audiences are gripped by the horrors of great screen monsters in the form of squids and sharks. Included here for comparison is a selection of real sea monsters—both modern and prehistoric.

Comparative sizes
All the giant marine creatures (1–9) on these two pages have been drawn to the same scale. A man in a 17ft-long kayak has also been drawn to scale to give a quick indication of actual size.

Modern marine monsters
Each of these creatures is the largest known recorded example of its kind.
1 *Lineus longissimus*, a ribbon worm—longest of all animals, 180ft.
2 The jellyfish *Cyanaea arctica* has 120ft tentacles.

3 Remains of an octopus suggest it had 100ft tentacles.
4 Whale shark, 60ft 9in.
5 Giant squid, 57ft long.
6 Starfish, *Midgardia xandaros*, 4ft 6in span.
7 Giant spider crab, with a claw span of 12ft 1½in.

8 Loggerhead sponge, 3ft 6in high, 3ft diameter.
9 American lobster, 3ft from tail to claw tip.

It would take a long time to wash with Leucosolenia blanca, *the smallest known sponge. Less than $\frac{1}{8}$ of an inch tall when fully grown, 30 of them would fit end to end across the palm of a man's hand.*

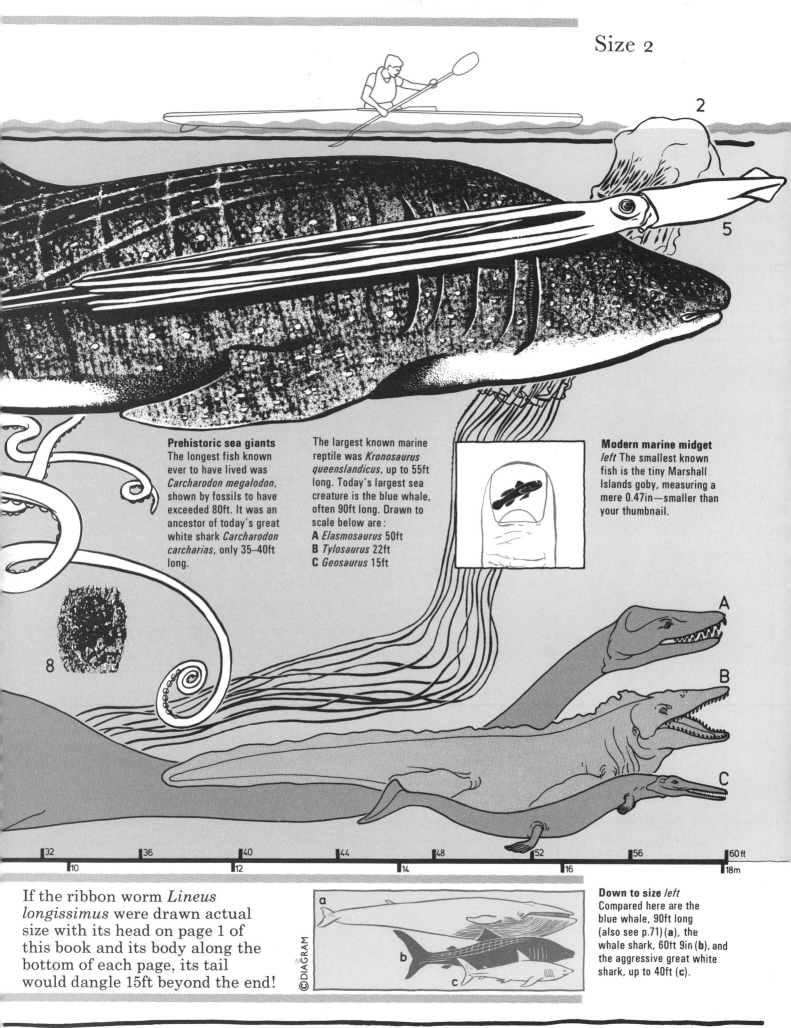

2

5

Prehistoric sea giants
The longest fish known ever to have lived was *Carcharodon megalodon*, shown by fossils to have exceeded 80ft. It was an ancestor of today's great white shark *Carcharodon carcharias*, only 35–40ft long.

The largest known marine reptile was *Kronosaurus queenslandicus*, up to 55ft long. Today's largest sea creature is the blue whale, often 90ft long. Drawn to scale below are:
A *Elasmosaurus* 50ft
B *Tylosaurus* 22ft
C *Geosaurus* 15ft

Modern marine midget
left The smallest known fish is the tiny Marshall Islands goby, measuring a mere 0.47in—smaller than your thumbnail.

8

A

B

C

32 36 40 44 48 52 56 60 ft
10 12 14 16 18m

If the ribbon worm *Lineus longissimus* were drawn actual size with its head on page 1 of this book and its body along the bottom of each page, its tail would dangle 15ft beyond the end!

©DIAGRAM

a
b
c

Down to size *left*
Compared here are the blue whale, 90ft long (also see p.71) (**a**), the whale shark, 60ft 9in (**b**), and the aggressive great white shark, up to 40ft (**c**).

REPTILES

The largest creatures ever to have roamed on Earth were probably reptiles—the huge dinosaurs of prehistoric times. *Diplodocus*, the longest dinosaur, makes today's largest lizard, the rare Komodo dragon, look like small fry. Today's largest crocodilian, the estuarine or saltwater crocodile, is certainly a force to be reckoned with, but at 20ft long it is much smaller than its 50ft-long prehistoric forebears. The biggest prehistoric turtle was probably about twice the length of the modern Pacific leatherback, but the largest prehistoric snakes are thought to have been little longer than modern anacondas and pythons.

Snakes and ladders
left, below A selection of present-day snakes— from the largest to the smallest—is illustrated here. Though often very large, constrictors, which kill by suffocation, are generally less dangerous than smaller venomous types.

A Anaconda, a constrictor from S America. Up to 30ft long, sometimes more.
B Reticulated python, a constrictor from SE Asia. May exceed 30ft.
C King cobra, from India. World's largest venomous snake, 18ft.
D Boa constrictor, from tropical America, 14ft.
E Eastern diamond-backed rattlesnake, venomous snake from N America. Largest rattler, 7–8ft.
F Mamba, venomous, from C and S Africa, 7ft.
G Grass snake, harmless European snake, 3ft.
H Thread snake, W Indies, world's shortest, 4½in.

Smallest reptile *above*
The smallest known reptile species, *Sphaerodactylus parthenopion*, is a type of gecko lizard. As shown here real size, adults measure less than ¾in from snout to vent, with a tail of approximately similar length.

ft		5		10		15	
m	1	2	3	4	5	6

A crocodile's egg, about 3¼in long, is approximately the same size as a goose egg.

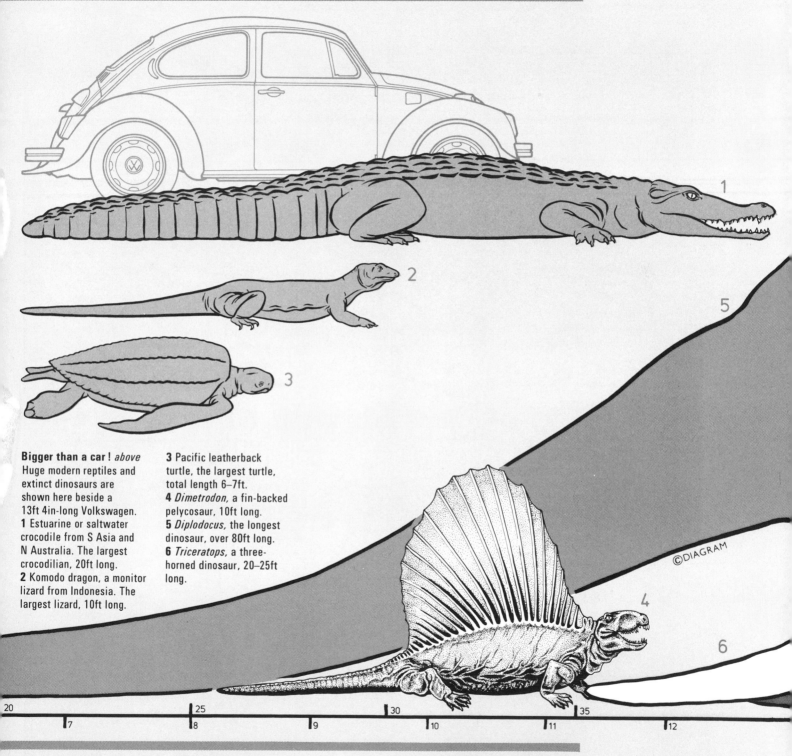

Bigger than a car! *above*
Huge modern reptiles and extinct dinosaurs are shown here beside a 13ft 4in-long Volkswagen.
1 Estuarine or saltwater crocodile from S Asia and N Australia. The largest crocodilian, 20ft long.
2 Komodo dragon, a monitor lizard from Indonesia. The largest lizard, 10ft long.

3 Pacific leatherback turtle, the largest turtle, total length 6–7ft.
4 *Dimetrodon,* a fin-backed pelycosaur, 10ft long.
5 *Diplodocus,* the longest dinosaur, over 80ft long.
6 *Triceratops,* a three-horned dinosaur, 20–25ft long.

©DIAGRAM

A chameleon can afford to "give you a piece of its tongue," for its tongue can be extended to a distance equal to its body length.

Turn over the page to see the bodies and heads of the huge *Diplodocus* (5) and *Triceratops* (6).

DINOSAURS

The huge dinosaurs or "terrible lizards" of Jurassic and Cretaceous times (190–65 million years ago) were the largest land creatures ever to have lived. Man did not appear for probably another 60 million years, but a man of average size (5ft 9in) is included here for comparison, drawn to the same scale as our six dinosaur examples.

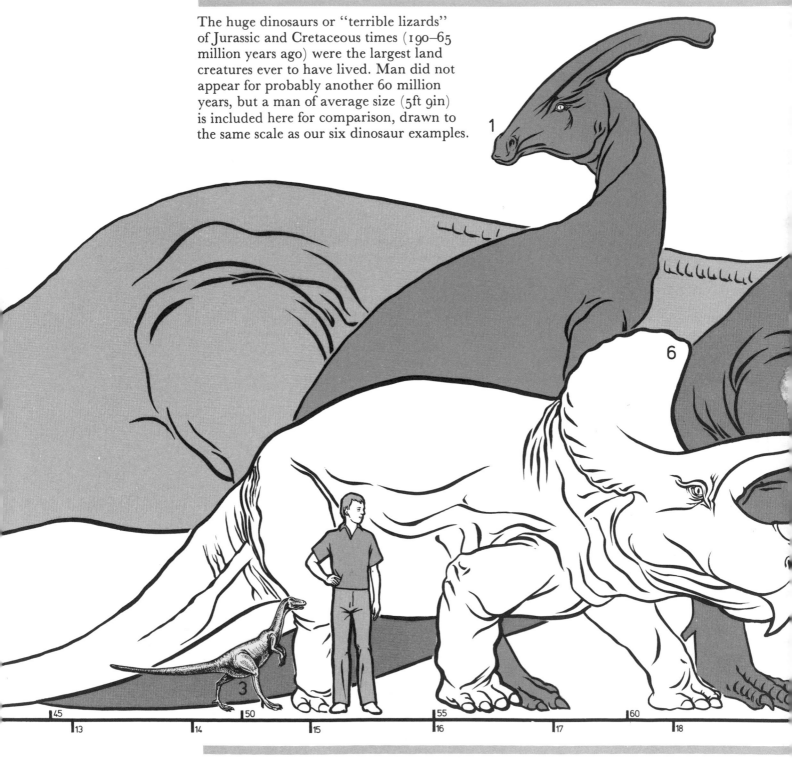

It's just possible that you could have jumped on the tail of a *Diplodocus* and got away with it— because of the time taken for a nerve impulse to travel from the tail to the brain and back again.

The Stegosaurus, *a playful 25ft long, had a brain the size of a walnut.*

1 *Parasaurolophus*, a duck-billed dinosaur with a large bony headcrest. A herbivore from late Cretaceous times, it was about 30ft long.
2 *Tyrannosaurus*, the largest carnivorous dinosaur. Remains from the Cretaceous period suggest it was 18ft high and perhaps up to 47ft long.

3 *Ornitholestes*, a small carnivorous dinosaur from Jurassic times. Its name means bird-catcher but it probably ate mainly reptiles. About 6ft long.
4 *Stegosaurus,* a heavily armored herbivore with big triangular bony plates like sails along its spine. From the Jurassic period, 12–13ft tall, 18–25ft long.
5 *Diplodocus,* a huge plant-eater from the Cretaceous period. The longest of the dinosaurs; a composite skeleton in Pittsburgh is 87½ft long.
6 *Triceratops*, a late Cretaceous dinosaur with three horns and a bony neck frill. A strong and agile fighter, 8–10ft tall and 20–25ft long.

©DIAGRAM

Some horned dinosaurs had horns up to 3ft long, making them longer than the legs of an average man.

INVERTEBRATES

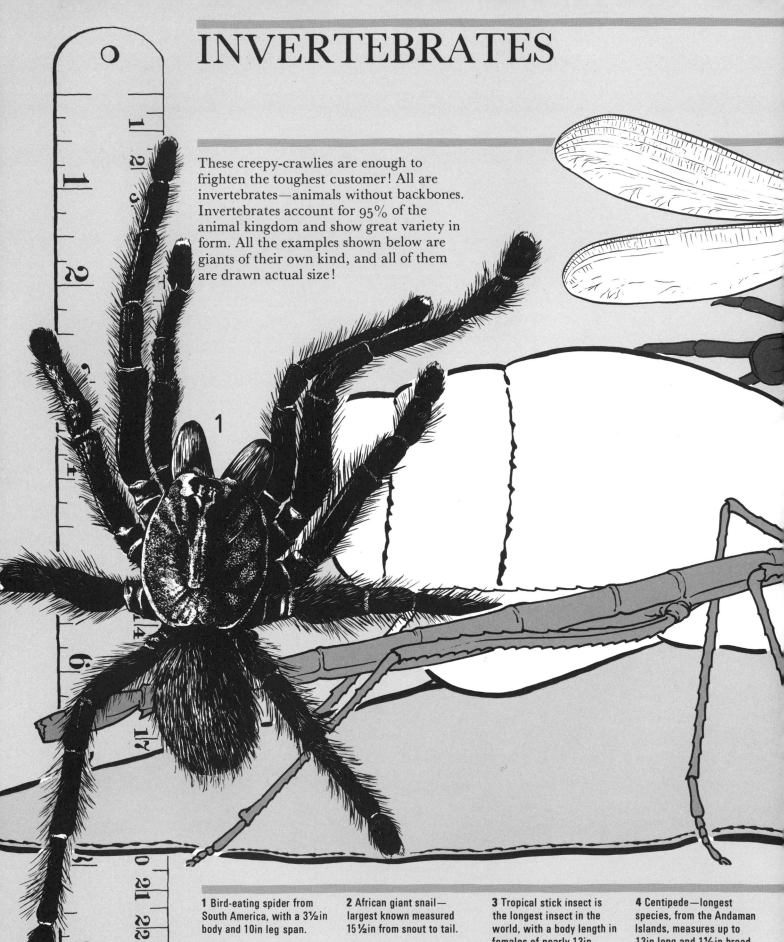

These creepy-crawlies are enough to frighten the toughest customer! All are invertebrates—animals without backbones. Invertebrates account for 95% of the animal kingdom and show great variety in form. All the examples shown below are giants of their own kind, and all of them are drawn actual size!

1

1 Bird-eating spider from South America, with a 3½in body and 10in leg span.

2 African giant snail—largest known measured 15½in from snout to tail.

3 Tropical stick insect is the longest insect in the world, with a body length in females of nearly 13in.

4 Centipede—longest species, from the Andaman Islands, measures up to 13in long and 1½in broad.

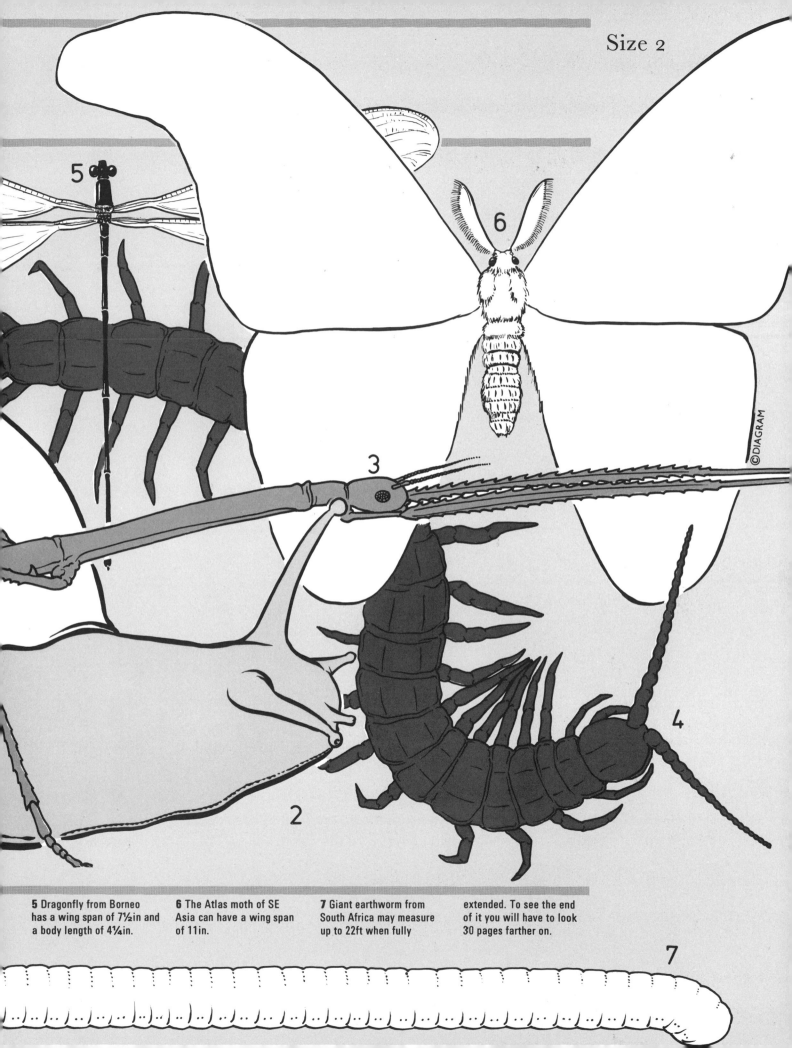

5

6

3

2

4

©DIAGRAM

5 Dragonfly from Borneo has a wing span of 7½in and a body length of 4¼in.

6 The Atlas moth of SE Asia can have a wing span of 11in.

7 Giant earthworm from South Africa may measure up to 22ft when fully extended. To see the end of it you will have to look 30 pages farther on.

7

BIRDS

Fossils of birdlike creatures suggest that the early birds caught the records for size. In fact these huge creatures were not birds at all, but pterosaurs. The largest known prehistoric bird was the elephant bird. Today the largest bird is the ostrich, but the wandering albatross has the greatest wing span.

Tiny flier
left The smallest bird in the world, shown here real size, is Helena's hummingbird from Cuba. An average adult male measures only 2¼in from bill tip to tail.

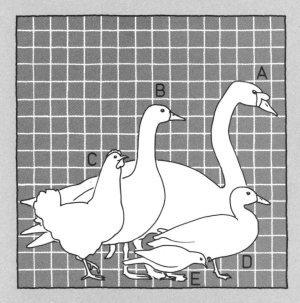

Domestic birds *above*
The swan (**A**), goose (**B**), hen (**C**), duck (**D**) and pigeon (**E**) have such different body proportions that it is easier to make a visual assessment of their comparative sizes than it is to express this in statistics.

Exceptional eggs *left*
The largest egg laid by any bird alive today is that of the ostrich, 6–8in long. The smallest is that of Helena's hummingbird, under ½in.

Flightless giants
above Flightless birds take the records for size. Our examples are shown to scale with a man of average height (5ft 9in).

1 North African ostrich, the tallest bird alive today. Males are sometimes up to 9ft tall.
2 Moa from New Zealand, probably the tallest extinct bird. May have been 13ft tall.

3 Elephant bird from Madagascar, thought to have been the most massive but not the tallest prehistoric bird, 9–10ft tall.

The kalong, a huge fruit bat from Indonesia, has a wing span of up to 5ft 7in—or roughly 1½ times the length of a baseball bat (3ft 6in).

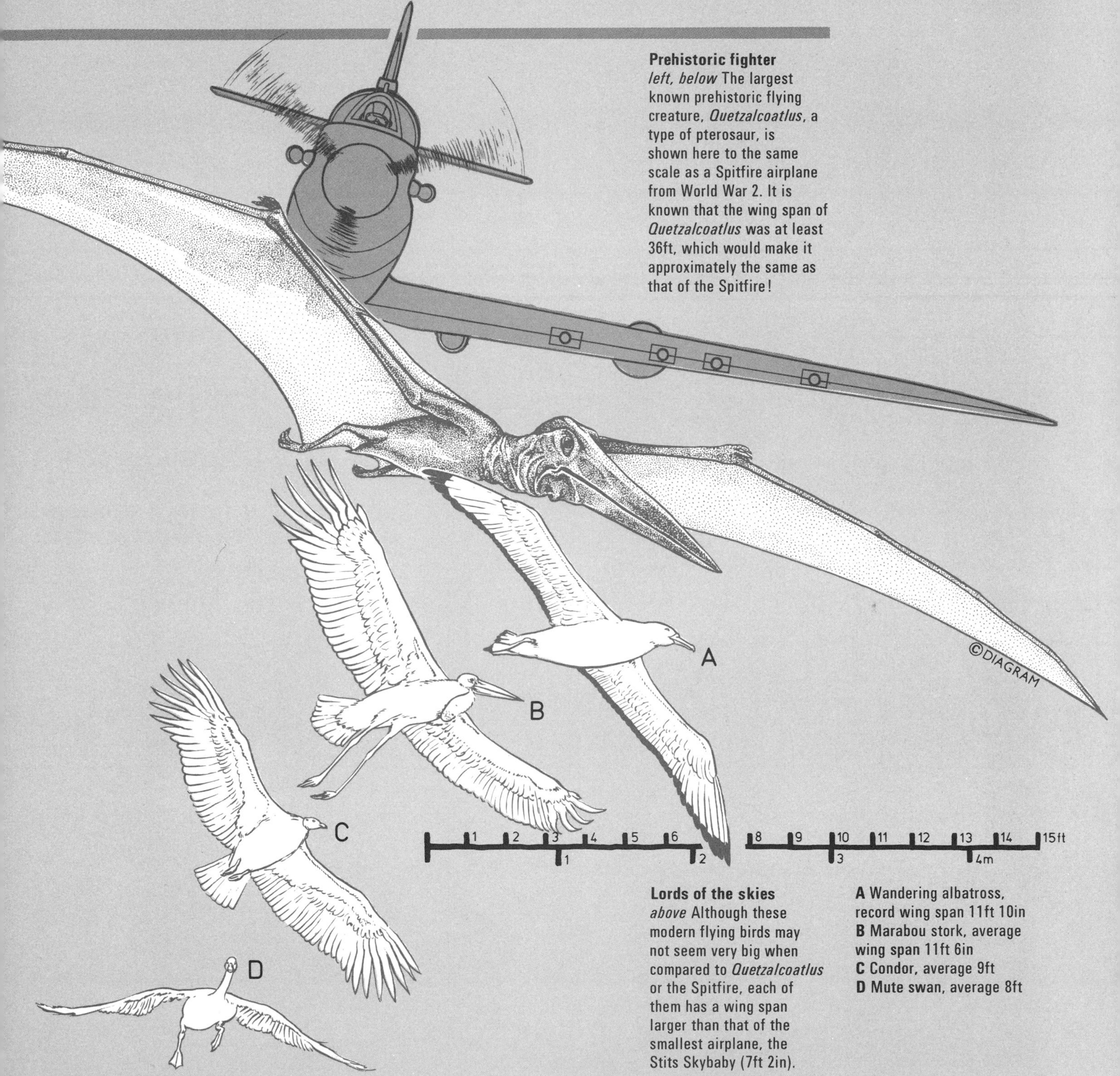

Prehistoric fighter
left, below The largest known prehistoric flying creature, *Quetzalcoatlus*, a type of pterosaur, is shown here to the same scale as a Spitfire airplane from World War 2. It is known that the wing span of *Quetzalcoatlus* was at least 36ft, which would make it approximately the same as that of the Spitfire!

©DIAGRAM

A

B

C

D

Lords of the skies
above Although these modern flying birds may not seem very big when compared to *Quetzalcoatlus* or the Spitfire, each of them has a wing span larger than that of the smallest airplane, the Stits Skybaby (7ft 2in).

A Wandering albatross, record wing span 11ft 10in
B Marabou stork, average wing span 11ft 6in
C Condor, average 9ft
D Mute swan, average 8ft

The 11ft 10in wing span of a wandering albatross is approximately twice the arm span of a man of average height.

MAMMALS

Today's largest mammal, the blue whale, is probably the largest mammal ever to have lived. The record for the largest land mammal, however, goes to a prehistoric creature, the *Paraceratherium*, a huge, hornless rhinolike creature that lived some 40 to 20 million years ago. The largest modern land animal is the African elephant. The tallest is the giraffe. Some of today's mammals, such as the tiger, rhino and armadillo, are smaller than their prehistoric forebears. Others, notably the horse, are very much bigger. All the extinct mammals below and the living mammals facing them are drawn to the same scale for easy visual comparison.

Extinct land mammals
A *Megatherium*, a type of sloth, about 20ft long.
B *Paraceratherium*, 35–37ft long, 18ft at shoulder.

C Mammoth, largest perhaps 14ft 9in at shoulder.
D Saber-toothed tiger, about 3ft 4in at shoulder.
E *Oxydactylus*, a Miocene camel, 4ft 6in at shoulder.

F *Synthetoceras*, 6ft long.
G *Glyptodon*, 9ft long.
H *Eohippus*, an early horse, 18in long, 9in at shoulder.

Eohippus, a prehistoric horse, was approximately the same size as a modern cat.

The mighty ape in the 1933 film "King Kong" appeared to be a terrifying 50ft tall—in reality, the model used was a mere 18in from head to toe.

Monster whale
left The blue whale is the largest of all mammals. Specimens 90ft long are not unusual; the longest ever recorded was over 110ft. To give an indication of comparative size, the illustration also includes an elephant and a giraffe.

Scale: 6 / 10 12 / 20 30 18 / 40 50 24 / 60 70 30 m / 80 90 100ft

Modern land mammals
1 Giraffe may tower 19ft above ground.
2 African elephant, male about 10ft 6in at shoulder.
3 Horse, usually 5ft to 5ft 6in at shoulder. Record 6ft 6in (19.2 hands).
4 Average man 5ft 9in tall; average woman 5ft 3¾in.

5 White rhino, 6ft 6in at shoulder, 16ft long.
6 Cow, 5ft at shoulder.
7 Bactrian camel, 7ft to top of humps.
8 Brown bear may be up to 8ft when standing erect.
9 Armadillo, total length 2ft 6in.
10 Cat, 9in at shoulder.

m ft
6 / 20
19
5 / 18
17
16

The tiny shrew
above Drawn actual size, the pygmy shrew has a head and body length of 1.7in. Its tail adds another 1.2in.

©DIAGRAM

A mammoth's shoulder height was a little greater than the height of a London double-decker bus (14ft 4½in); *Paraceratherium* could have looked over such a bus with ease.

HUMANS

In reality there are no such persons as Mr and Mrs Average, but studies of the vital statistics of US adult males and females provide some interesting comparisons. For example, although the height range of the "normal" (95% of all males or females) is quite small (only 10in), the range of the possible is surprisingly great (about 6ft).

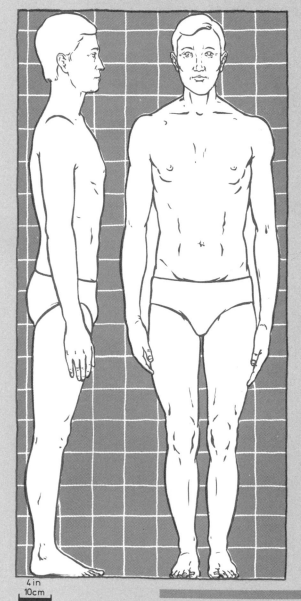

4 in
10 cm

Mr Average
left The dimensions of the "average" US male are as follows:
Height 5ft 9in
Weight 162lb
Chest 38¾in
Waist 31¾in
Hips 37¾in

Large and little
right The "average" US male is shown here, plus figures representing the upper and lower limits of growth—both "normal" (95% of the adult male population) and extreme.
A The tallest reliably measured male was Robert Wadlow (1918–40) of the USA, 8ft 11in.
B Upper limit of normal range, 6ft 2in.
C Average US male, 5ft 9in.
D Lower limit of normal range, 5ft 4in.
E The shortest recorded adult male was Calvin Phillips (1791–1812) of the USA, who measured 2ft 2½in.

While the head is one quarter of the total length at birth, it is only one sixth by the age of six and one eighth by adulthood.

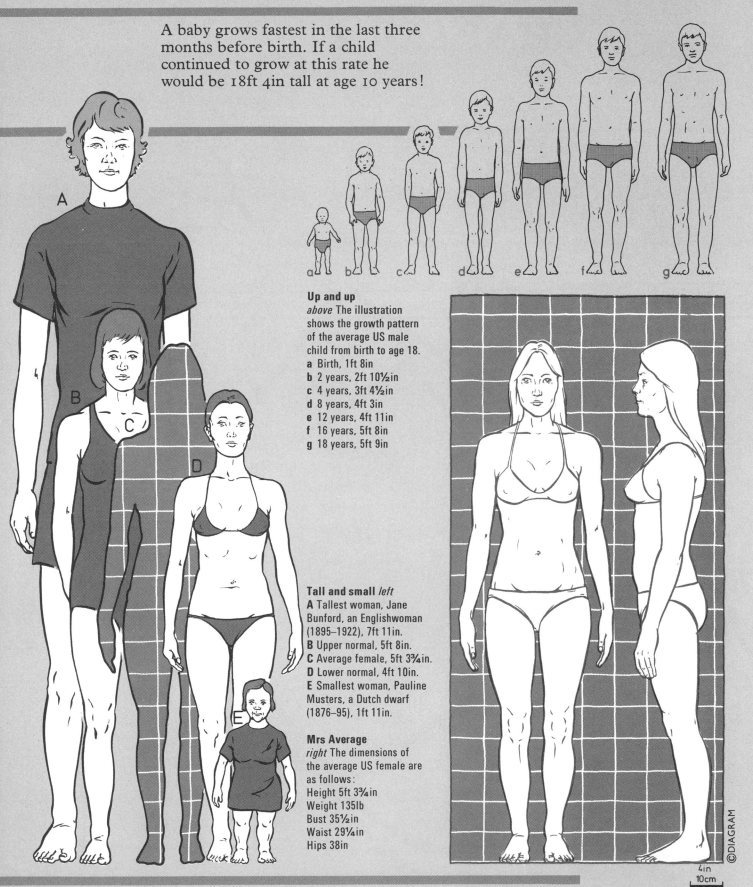

A baby grows fastest in the last three months before birth. If a child continued to grow at this rate he would be 18ft 4in tall at age 10 years!

Up and up

above The illustration shows the growth pattern of the average US male child from birth to age 18.

a Birth, 1ft 8in
b 2 years, 2ft 10½in
c 4 years, 3ft 4½in
d 8 years, 4ft 3in
e 12 years, 4ft 11in
f 16 years, 5ft 8in
g 18 years, 5ft 9in

Tall and small *left*
A Tallest woman, Jane Bunford, an Englishwoman (1895–1922), 7ft 11in.
B Upper normal, 5ft 8in.
C Average female, 5ft 3¾in.
D Lower normal, 4ft 10in.
E Smallest woman, Pauline Musters, a Dutch dwarf (1876–95), 1ft 11in.

Mrs Average
right The dimensions of the average US female are as follows:
Height 5ft 3¾in
Weight 135lb
Bust 35½in
Waist 29¼in
Hips 38in

©DIAGRAM

4in
10cm

The average male is taller than the average female at all ages except around age 12 years, when the girl's pre-puberty growth spurt puts her briefly ahead.

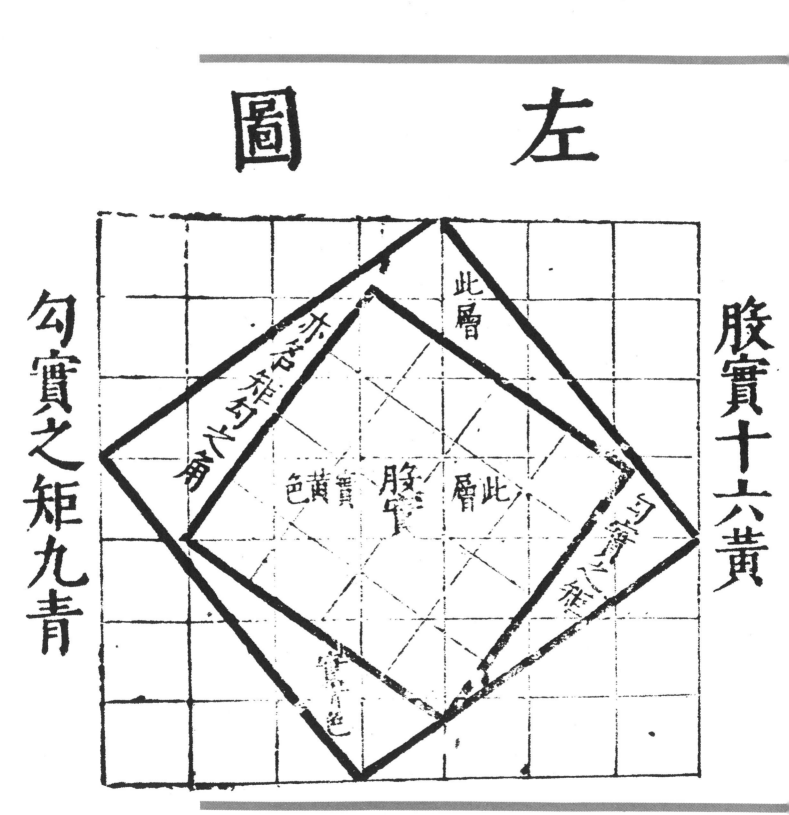

CHAPTER 4

This illustration of what is now generally known as Pythagoras' theorem is from *Ch'ou-pei Suan-king*, an ancient Chinese treatise dating from c. 1100BC (British Museum).

AREA AND VOLUME

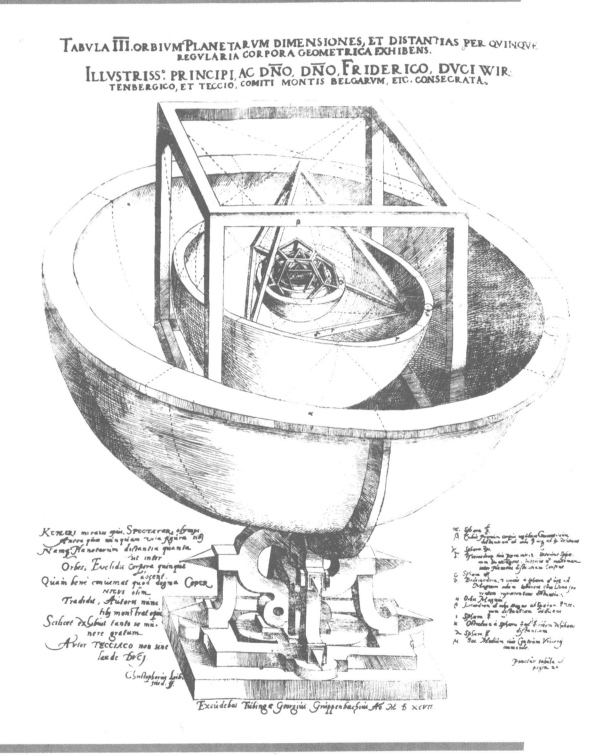

From Kepler's *Continens Misterium Cosmografica* of 1596, this diagram shows models of the five perfect solids fitted between spheres representing the orbits of the six planets known at that time (Science Museum, London).

MEASURING AREA

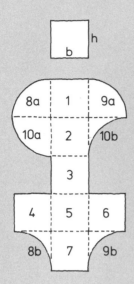

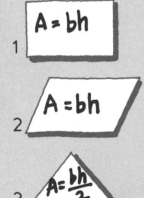

Area measurements describe the size of surfaces, which may be either flat or curved, in terms of the number of square units that can be fitted within them. The sporadic development of the US/imperial units has resulted in a variety of multiplication factors, whereas the metric system is firmly based on multiples of 10.

Basic US customary/imperial units of area
144 square inches (in^2, sq.in) = 1 square foot (ft^2, sq.ft)
9 square feet = 1 square yard (yd^2, sq.yd)
4840 square yards = 1 acre (a, ac)
640 acres = 1 square mile (mi^2, sq.mi)
Other US/imperial units of area
1 square rod, pole or perch = $30\frac{1}{4}yd^2$
1 square chain = 16 square rods = $484yd^2$
1 rood = $2\frac{1}{2}$ square chains = $\frac{1}{4}$ acre = $1210yd^2$

Measuring area *left*
The area of a square is its base (**b**) multiplied by its height (**h**). Shown is a square with sides of 1cm. Using this square centimeter (cm^2) as a unit of measurement we see that the irregular shape shown has an area of $10cm^2$.

Metric area units *right*
The table gives the basic units. Square millimeters, centimeters, meters and kilometers have been expressed by the standard abbreviations (mm^2, cm^2, m^2 and km^2). Not used here are the abbreviations for are (a) and hectare (ha).

Metric units of area
$100mm^2$	= $1cm^2$
$10,000cm^2$	= $1m^2$
$100m^2$	= 1 are
100 ares	= 1 hectare
100 hectares	= $1km^2$

US/imperial units of area
above Included here are basic and also less common units of area from the US and imperial systems. Given in brackets are alternative abbreviations in common use; the abbreviations given first, for example in^2, are the ones used in this book.

Geometric areas *left*
Written inside these common geometric shapes are the formulae for calculating their areas.
Shapes shown are:
1 Rectangle
2 Parallelogram
3 Triangle
4 Trapezium
5 Circle
Abbreviations used are:
A = area
a = top
b = base
h = height
π = 3.1416
r = radius

Visualizing areas *below*
All these areas are drawn to a common scale.
A A square chain ($484yd^2$) is shown together with a basketball court ($427yd^2$).
B An acre ($4840yd^2$) is a little larger than two ice hockey rinks ($2 \times 2222yd^2$).
C An are ($119.6yd^2$) is shown with a boxing ring ($44yd^2$).

D A hectare ($11,960yd^2$) is about $\frac{1}{3}$ as big again as a soccer pitch ($8800yd^2$).

1 $A = bh$

2 $A = bh$

3 $A = \dfrac{bh}{2}$

4 $A = \dfrac{(a+b)h}{2}$

5 $A = \pi r^2$

A

B

C

D

1 square foot = 1.6 pages of this book.
1 square yard = 14.3 pages of this book.
1 square meter = 17.1 pages of this book.

Square measures *right*
Shown here real size for easy visual comparison are a square inch (1) and a square centimeter (2).

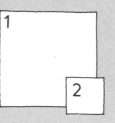

Conversion tables *left*
To convert US/imperial area measurements into metric, and vice versa, find the figure to be converted—US/imperial or metric—in the central column. Its equivalent will be in the appropriate column to the right or the left.

in²		cm²	yd²		m²	acre		ha	mi²		km²
0.1550	1	6.4516	1.1960	1	0.8361	2.4710	1	0.4047	0.3861	1	2.5900
0.3100	2	12.903	2.3920	2	1.6722	4.9421	2	0.8094	0.7722	2	5.1800
0.4650	3	19.355	3.5880	3	2.5084	7.4131	3	1.2141	1.1583	3	7.7699
0.6200	4	25.806	4.7840	4	3.3445	9.8842	4	1.6187	1.5444	4	10.360
0.7750	5	32.258	5.9800	5	4.1806	12.355	5	2.0234	1.9305	5	12.950
0.9300	6	38.710	7.1759	6	5.0168	14.826	6	2.4281	2.3166	6	15.540
1.0850	7	45.161	8.3719	7	5.8529	17.297	7	2.8328	2.7027	7	18.130
1.2400	8	51.613	9.5679	8	6.6890	19.768	8	3.2375	3.0888	8	20.720
1.3950	9	58.064	10.764	9	7.5251	22.239	9	3.6422	3.4749	9	23.310
2.3500	15	96.774	17.940	15	12.542	37.066	15	6.0703	5.7915	15	38.850
3.8750	25	161.29	29.900	25	20.903	61.776	25	10.117	9.6525	25	64.750
5.4250	35	225.81	41.860	35	29.264	86.487	35	14.164	13.514	35	90.650
6.9750	45	290.32	53.820	45	37.626	111.20	45	18.211	17.375	45	116.55
8.5250	55	354.84	65.779	55	45.987	135.91	55	22.258	21.236	55	142.45
10.075	65	419.35	77.739	65	54.348	160.62	65	26.305	25.097	65	168.35
11.625	75	483.87	89.699	75	62.710	172.97	75	30.351	28.958	75	194.25
13.175	85	548.39	101.66	85	71.071	210.04	85	34.398	32.819	85	220.15
14.725	95	612.90	113.62	95	79.432	234.75	95	38.445	36.680	95	246.05

©DIAGRAM

COMPARATIVE AREAS

Comparisons of area—a two-dimensional property involving both length and breadth—are generally harder to grasp than are comparisons based on length alone. Here we use metric measurements and multiples of 10 to explore areas from a microscopic square as small as a virus to a square with an area as big as our solar system.

An area concertina
Different areas are here represented by a series of squares of constant size. We start with a colored square that is an enlargement of the 1cm² drawn actual size beside it. Moving left from the starting square, each large square has an area one tenth of the preceding one. Moving right, each square has an area 10 times greater than the one before. The effect of increasing or decreasing areas in this way is also demonstrated here by our series of illustrated examples.

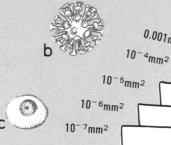

1cm²

10 mm²

1mm²

0.1mm²

0.01mm²

0.001mm²

10⁻⁴mm²

10⁻⁵mm²

10⁻⁶mm²

10⁻⁷mm²

10⁻⁸mm²

10cm²

100cm²

1000cm²

1m²

10m²

100m²

1000m²

a
b
c

A
B
C
D

Going down in area
a The top of a flat pinhead has an area of approximately 1mm².
b *Discosphaera*, a type of plankton with a cross-sectional diameter of about 0.03mm, would fit on a square with an area of 0.001mm².
c *Escherichia coli*, a tiny bacterium, has a diameter of 0.001mm, equal to the sides of a square with an area of 10⁻⁶mm².
d An influenza virus, measuring 0.0001mm across, will fit on a square with an area of 10⁻⁸mm².

100,000,000,000 bacteria the size of *Escherichia coli* could fit on a food tray measuring 40x25cm.

The total skin area of an average man (20ft²) and an average woman (17ft²) can easily be covered by a sheet for a single bed (48ft²).

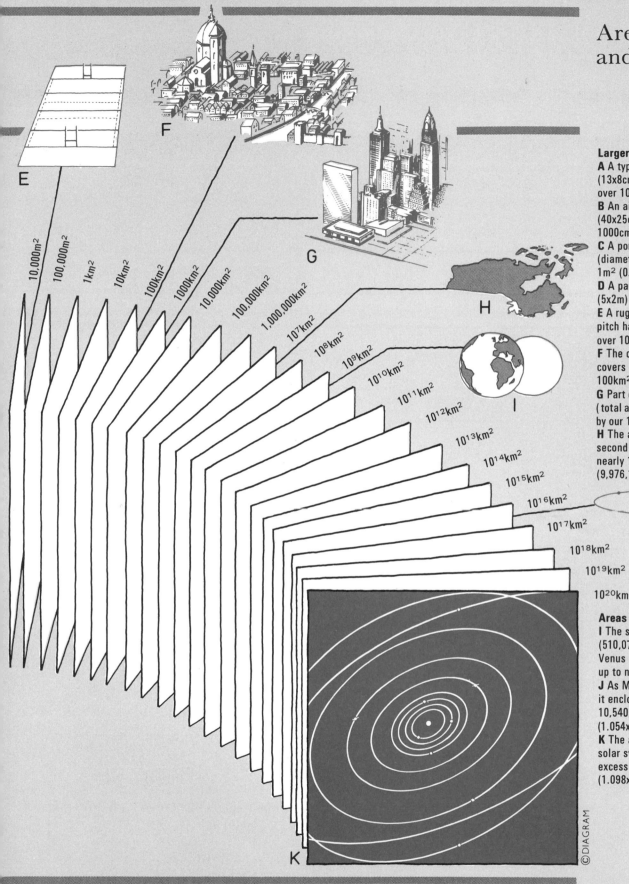

Area and volume

Larger and larger areas

A A typical pocket diary (13x8cm) has an area just over 100cm² (104cm²).

B An airline food tray (40x25cm) has an area of 1000cm².

C A portable paddling pool (diameter 110cm) is nearly 1m² (0.95m²).

D A parking space for a car (5x2m) measures 10m².

E A rugby union football pitch has an area slightly over 10,000m² (10,039m²).

F The city of Florence, Italy, covers an area just over 100km² (102km²).

G Part of New York City (total area 830km²) is shown by our 1000km² square.

H The area of Canada, the second largest country, is nearly 10⁷km² (9,976,139km²).

Areas in space

I The surface areas of Earth (510,070,000km²) and Venus (460,270,000km²) add up to nearly 10⁹km².

J As Mercury orbits the Sun it encloses an area of 10,540,000,000,000,000km² (1.054x10¹⁶km²).

K The area in which our solar system operates is in excess of 10²⁰km² (1.098x10²⁰km²).

10,000m²
100,000m²
1km²
10km²
100km²
1000km²
10,000km²
100,000km²
1,000,000km²
10⁷km²
10⁸km²
10⁹km²
10¹⁰km²
10¹¹km²
10¹²km²
10¹³km²
10¹⁴km²
10¹⁵km²
10¹⁶km²
10¹⁷km²
10¹⁸km²
10¹⁹km²
10²⁰km²

©DIAGRAM

If Central Park, New York City were transformed into a vast parking lot, there would be space for over 300,000 cars.

One page of this book has a surface area of 585.48cm² or 90.75in².

THE PLANETS

The diameter of Jupiter, the largest planet, is nearly 30 times as big as that of Mercury, the smallest planet, but its estimated surface area is over 850 times as big. On Earth's surface there is over twice as much sea as land, and one ocean, the Pacific, has an area over six million square miles bigger than the entire land surface.

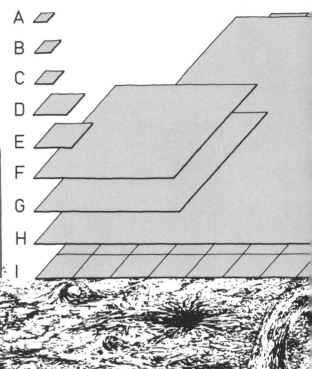

Surface area of planets

Listed in the table *right* and shown to a common scale as squares *far right* are estimated surface areas of the planets. The small squares on square I show how many times Earth's surface area will fit into that of Jupiter (125).

		mi²	km²
A	Mercury	28,880,000mi²	74,800,000km²
B	Pluto	43,000,000mi²	111,000,000km²
C	Mars	55,870,000mi²	144,700,000km²
D	Venus	177,710,000mi²	460,270,000km²
E	Earth	196,940,000mi²	510,070,000km²
F	Neptune	2,980,200,000mi²	7,718,700,000km²
G	Uranus	3,257,300,000mi²	8,436,400,000km²
H	Saturn	17,483,000,000mi²	45,281,000,000km²
I	Jupiter	24,717,000,000mi²	64,017,000,000km²

Area of continents

Listed in the table *right* and shown to scale in the diagram *far right* are the areas of the continents. Asia is 795,000mi² larger than N and S America combined. Asia and Africa together make up just over half the total land area.

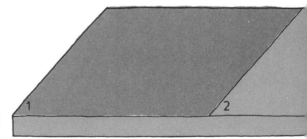

		mi²	km²
1	Asia	17,085,000mi²	44,250,000km²
2	Africa	11,685,000mi²	30,264,000km²
3	N America	9,420,000mi²	24,398,000km²
4	S America	6,870,000mi²	17,793,000km²
5	Antarctica	5,100,000mi²	13,209,000km²
6	Europe	3,825,000mi²	9,907,000km²
7	Oceania	3,295,000mi²	8,534,000km²

Oceans and seas

Listed *right* and shown to scale in the diagram *far right* are the world's oceans and five largest seas. The four oceans combined account for approximately 92% of the total sea area, with the Pacific alone accounting for about 46%.

		mi²	km²
a	Pacific Ocean	63,800,000mi²	165,242,000km²
b	Atlantic Ocean	31,800,000mi²	82,362,000km²
c	Indian Ocean	28,400,000mi²	73,556,000km²
d	Arctic Ocean	5,400,000mi²	13,986,000km²
e	Malay Sea	3,144,000mi²	8,143,000km²
f	Caribbean Sea	1,063,000mi²	2,753,000km²
g	Mediterranean Sea	967,000mi²	2,505,000km²
h	Bering Sea	876,000mi²	2,269,000km²
i	Gulf of Mexico	596,000mi²	1,544,000km²

If the world's total land area were shared equally among the world's population, each person would have a plot of 8.5 acres. Climate and terrain, however, mean that perhaps 80% of these plots would be useless.

The surface area of the smallest planet in our solar system, Mercury (28,880,000mi²), is approximately equal to the combined areas of Asia and Africa (28,770,000mi²).

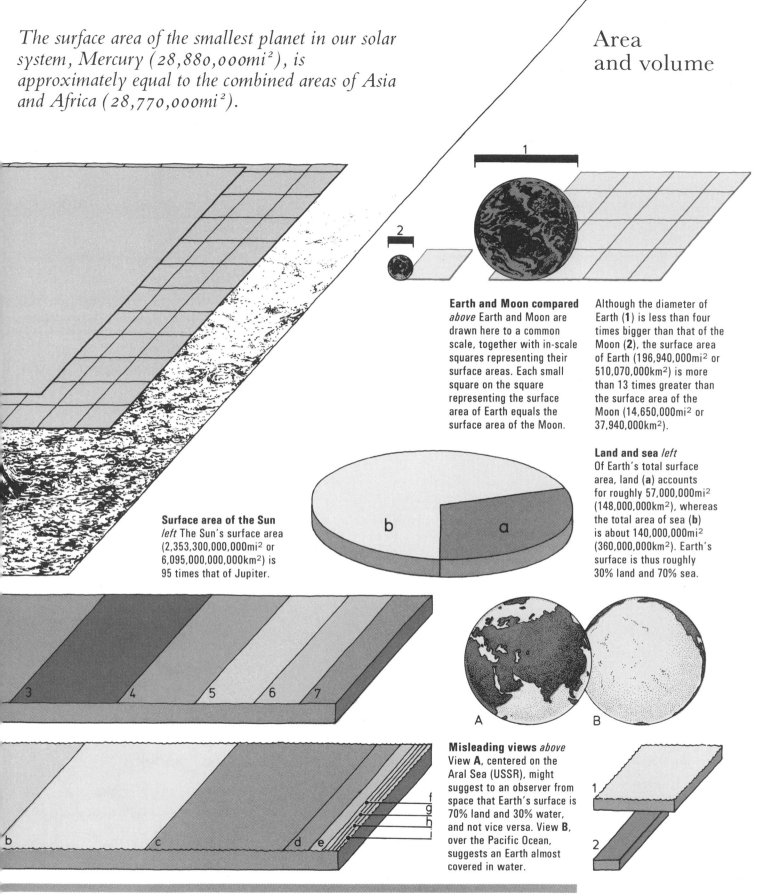

Earth and Moon compared *above* Earth and Moon are drawn here to a common scale, together with in-scale squares representing their surface areas. Each small square on the square representing the surface area of Earth equals the surface area of the Moon.

Although the diameter of Earth (**1**) is less than four times bigger than that of the Moon (**2**), the surface area of Earth (196,940,000mi² or 510,070,000km²) is more than 13 times greater than the surface area of the Moon (14,650,000mi² or 37,940,000km²).

Surface area of the Sun *left* The Sun's surface area (2,353,300,000,000mi² or 6,095,000,000,000km²) is 95 times that of Jupiter.

Land and sea *left* Of Earth's total surface area, land (**a**) accounts for roughly 57,000,000mi² (148,000,000km²), whereas the total area of sea (**b**) is about 140,000,000mi² (360,000,000km²). Earth's surface is thus roughly 30% land and 70% sea.

Misleading views *above* View **A**, centered on the Aral Sea (USSR), might suggest to an observer from space that Earth's surface is 70% land and 30% water, and not vice versa. View **B**, over the Pacific Ocean, suggests an Earth almost covered in water.

An equal distribution of the world's seas among the world's population would give each person an area approximately equal to 80 Olympic swimming pools. An equal share of the land is roughly equal to four and a half soccer pitches.

World's largest *above* The area of the largest ocean (**1**)—the Pacific at 63,800,000mi²—is over three times that of the largest continent (**2**)—Asia at 17,085,000mi².

81

LAKES AND ISLANDS

Here we compare the areas of the world's largest lakes and islands. North America's Great Lakes include the world's largest freshwater lake (Superior) and three more from the top 10 (Huron, Michigan, Erie). Greenland, the largest island, is larger than the next three largest put together (New Guinea, Borneo, Madagascar).

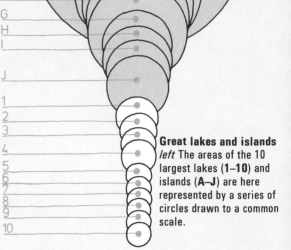

Great lakes and islands
left The areas of the 10 largest lakes (**1–10**) and islands (**A–J**) are here represented by a series of circles drawn to a common scale.

World's largest lakes
The world's 10 largest freshwater lakes are listed *right*, located on the map *below* and drawn to an equal area projection *below right*. The largest saltwater lake, the Caspian Sea (USSR), has an area of 143,550mi².

1	Superior	31,820mi²	82,414km²
2	Victoria	26,828mi²	69,485km²
3	Huron	23,010mi²	59,596km²
4	Michigan	22,400mi²	58,016km²
5	Great Bear	12,275mi²	31,792km²
6	Baykal	12,159mi²	31,492km²
7	Great Slave	10,980mi²	28,438km²
8	Tanganyika	10,965mi²	28,399km²
9	Malawi	10,900mi²	28,231km²
10	Erie	9,940mi²	25,745km²

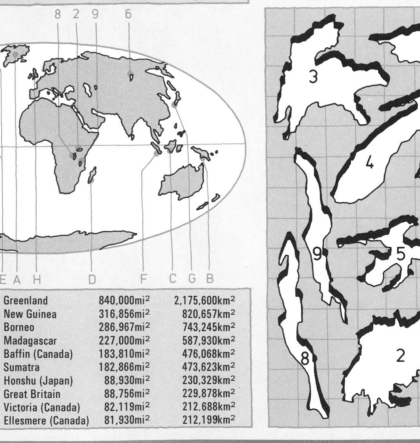

World's largest islands
The 10 largest are listed *right* and located *above*. Australia (2,941,526mi²) is not included, being classed as a continental land mass. Drawn with the islands to an equal area projection *far right* is the USA (Alaska/Hawaii excepted).

A	Greenland	840,000mi²	2,175,600km²
B	New Guinea	316,856mi²	820,657km²
C	Borneo	286,967mi²	743,245km²
D	Madagascar	227,000mi²	587,930km²
E	Baffin (Canada)	183,810mi²	476,068km²
F	Sumatra	182,866mi²	473,623km²
G	Honshu (Japan)	88,930mi²	230,329km²
H	Great Britain	88,756mi²	229,878km²
I	Victoria (Canada)	82,119mi²	212,688km²
J	Ellesmere (Canada)	81,930mi²	212,199km²

The world's largest desert, the Sahara, with an area of 3,250,000mi², is over three times bigger than the Mediterranean Sea to its north.

If Lake Superior (31,820mi²) were to be drained, the area of regained land would be just over twice the area of the Netherlands (15,770mi²).

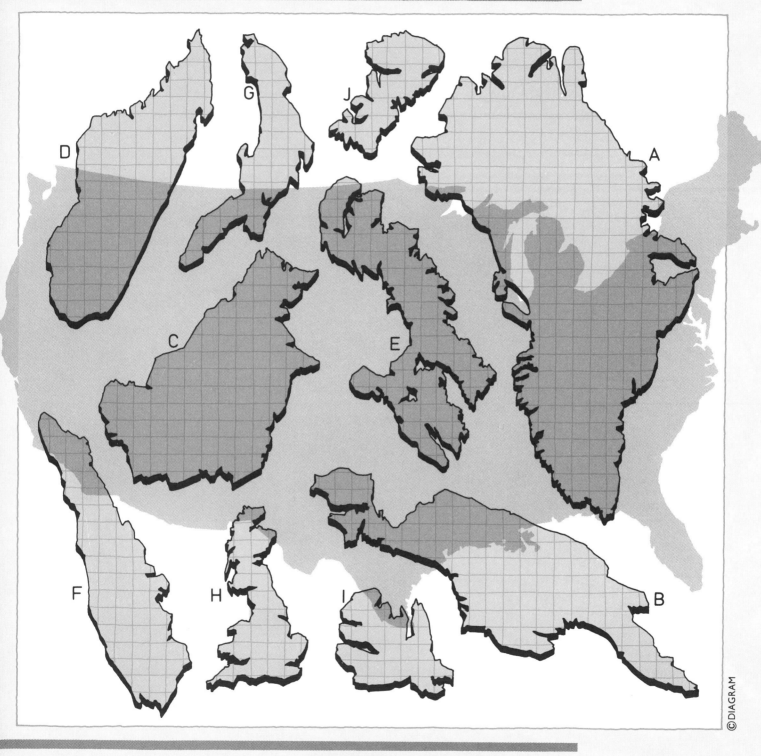

Lake Malawi (10,900mi²) is almost a quarter as big as the country of the same name (45,747mi²).

©DIAGRAM

COUNTRIES AND STATES 1

Different political and geographical factors have produced nation states of widely differing size. The smallest country in the world, the Vatican City State, is less than a quarter of a square mile in area, whereas the largest country in the world, the USSR has an area exceeding eight and a half million square miles.

Largest countries
The areas of the world's 10 largest countries are included in the list *below* and represented by the scaled squares beneath it. In the illustration *right* these same countries are drawn to an equal area projection, with the Moon drawn to the same scale in the center. Square **A** represents the land area of Earth (57,000,000mi²), and square **B** the surface of the Moon (14,650,000mi²). The USSR, with an area of 8,649,489mi², would cover well over half of the Moon's surface.

Smallest countries
right Listed here in ascending order of size are the 10 smallest countries in the world. Their combined areas (770mi²) would fit almost 1½ times into the state of Rhode Island, USA (area 1,049mi²).

A	Vatican City	0.17mi²	0.44km²
B	Monaco	0.73mi²	1.89km²
C	Nauru	8.1mi²	21km²
D	San Marino	23.4mi²	60.5km²
E	Liechtenstein	61.8mi²	160km²
F	Maldive Islands	115mi²	298km²
G	Malta	122mi²	316km²
H	Grenada	133mi²	344km²
I	St Vincent	150mi²	389km²
J	Seychelles	156mi²	404km²

1	USSR	8,649,489mi²	22,402,200km²
2	Canada	3,851,787mi²	9,976,139km²
3	China (CPR)	3,691,502mi²	9,561,000km²
4	USA	3,675,547mi²	9,519,666km²
5	Brazil	3,286,470mi²	8,511,965km²
6	Australia	2,966,136mi²	7,682,300km²
7	India	1,269,338mi²	3,287,590km²
8	Argentina	1,072,157mi²	2,776,889km²
9	Sudan	967,494mi²	2,505,813km²
10	Zaire	905,360mi²	2,344,885km²

Country parks?
The areas of the world's 10 smallest countries are represented *right* by 10 scaled squares. As shown *below*, New York City's Central Park (**a**) area 1.3mi², is nearly twice as big as the second-smallest country, Monaco (**b**) 0.73mi².

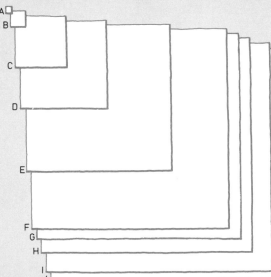

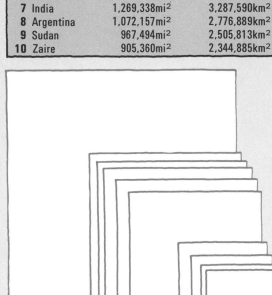

Bigger than Europe
right The USA (excluding Alaska and Hawaii) is here drawn to the same scale as Europe. The area of the entire USA (3,675,547mi²) exceeds by over one million square miles the area of all of Europe west of the USSR (2,572,600mi²).

Only six states?
far right The state of Alaska (586,412mi²) is here superimposed over the main area of the USA. If all the states were the size of Alaska there would be room for only six complete states in the entire area of the USA (3,675,547mi²).

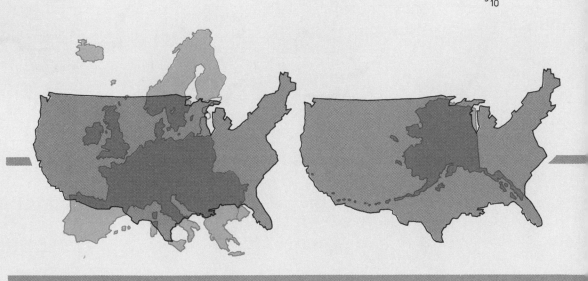

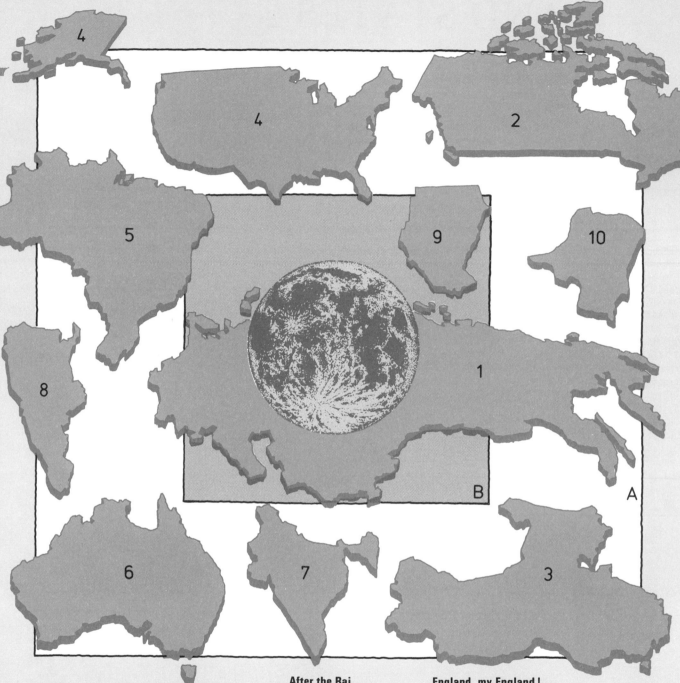

After the Raj
left India is the seventh largest country in the world, with an area of 1,269,338mi². The area of India's former ruler, the United Kingdom (93,026mi²), would fit over 13½ times into the area of the modern Indian state.

England, my England!
right New England (USA) is here drawn to scale over a map of England (UK). New England—Maine, New Hampshire, Vermont, Rhode Island, Massachusetts and Connecticut—at 62,992mi² is considerably larger than England at 50,053mi².

© DIAGRAM

COUNTRIES AND STATES 2

For its area, Europe has more countries than any other continent. Even France, Europe's largest country ($211,207mi^2$), ranks only forty-fourth in the world in terms of size. Canada's Northwest Territories ($1,304,900mi^2$) is six times bigger than France, and, if a country, would be 19th largest in the world.

1	France	$211,207mi^2$	$547,026km^2$
2	Spain	$194,896mi^2$	$504,782km^2$
3	Sweden	$173,731mi^2$	$449,964km^2$
4	Finland	$130,119mi^2$	$337,009km^2$
5	Norway	$125,181mi^2$	$324,219km^2$
6	Poland	$120,725mi^2$	$312,677km^2$
7	Italy	$116,316mi^2$	$301,260km^2$
8	Yugoslavia	$98,766mi^2$	$255,804km^2$
9	West Germany	$95,992mi^2$	$248,620km^2$
10	United Kingdom	$93,026mi^2$	$240,937km^2$
11	Romania	$91,699mi^2$	$237,500km^2$
12	Greece	$50,944mi^2$	$131,944km^2$
13	Czechoslovakia	$49,373mi^2$	$127,876km^2$
14	Bulgaria	$42,823mi^2$	$110,912km^2$
15	East Germany	$41,768mi^2$	$108,178km^2$
16	Iceland	$39,768mi^2$	$103,000km^2$
17	Hungary	$35,920mi^2$	$93,032km^2$
18	Portugal	$35,553mi^2$	$92,082km^2$
19	Austria	$32,374mi^2$	$83,850km^2$
20	Irish Republic	$27,136mi^2$	$70,282km^2$

Largest in Europe
The areas of the 20 largest countries in Europe are listed in the table *left*, drawn to an equal area projection *right* and represented as rectangles drawn to scale *below*. The remaining 12 countries that are completely in Europe have a combined area of just under $75,000mi^2$.

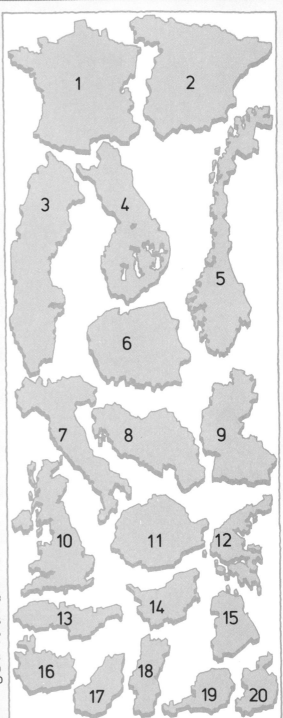

If Mexico ($761,604mi^2$) were to annex New Mexico, Texas and California, its new total area ($1,309,048mi^2$) would move it up from the thirteenth to the seventh largest country in the world.

In Iceland, the least densely populated country
in Europe, there is enough land for each person
to have 123.5 acres in an equal share-out; in Malta,
the most densely populated European country,
68 people would have to share each acre.

Common denominator
right France is the largest
country in the Common
Market (EEC), with an area
of 211,207mi². The diagram
shows the number of times
that each of the other eight
countries would fit into the
area of France.

Italy
(116,316mi², 301,260km²) x 1.8

West Germany
(95,992mi², 248,620km²) x 2.2

United Kingdom
(93,026mi², 240,937km²) x 2.3

Ireland
(27,136mi², 70,282km²) x 7.8

Denmark
(16,629mi², 43,069km²) x 12.7

Netherlands
(15,770mi², 40,844km²) x 13.4

Belgium
(11,781mi², 30,513km²) x 17.9

Luxembourg
(998mi², 2586km²) x 211.6

USSR: largest republics

1 Kazakhstan	1,049,100mi²	2,717,300km²
2 Ukraine	233,100mi²	603,700km²
3 Turkmenistan	188,500mi²	488,100km²
4 Uzbekistan	172,700mi²	447,400km²
5 Belorussia	80,200mi²	207,600km²

Canada: largest provinces/territories

1 Northwest Territories	1,304,900mi²	3,379,700km²
2 Quebec	594,900mi²	1,540,700km²
3 Ontario	412,600mi²	1,068,600km²
4 British Columbia	366,300mi²	948,600km²
5 Alberta	255,300mi²	661,200km²

China: largest administrative regions

1 Sian	1,260,100mi²	3,263,600km²
2 Chungking	926,900mi²	2,400,800km²
3 Shenyang	474,900mi²	1,230,000km²
4 Wuhan	392,600mi²	1,016,800km²
5 Peking	321,100mi²	831,600km²

USA: largest states

1 Alaska	586,400mi²	1,518,800km²
2 Texas	267,300mi²	692,300km²
3 California	158,700mi²	411,000km²
4 Montana	147,100mi²	381,000km²
5 New Mexico	121,400mi²	314,400km²

Political parts *left*
The table lists the areas of
the largest five political
subdivisions in each of the
four largest countries in the
world—the USSR, Canada,
China and the USA. Largest
of all is Canada's Northwest
Territories (1,304,900mi²).

Some of the parts *above*
The diagram shows the
comparative areas of the
largest land division in
each of the four largest
countries (in color), and
compares them with the
areas of countries that
are similar in size (in
white). See list *right*.

A Northwest Territories,
Canada (1,304,900mi²) and
India (1,269,338mi²).
B Sian, China (1,260,100mi²)
and India (1,269,338mi²).
C Kazakhstan, USSR
(1,049,100mi²) and
Argentina (1,072,157mi²).
D Alaska, USA (586,400mi²)
and Iran (636,293mi²).

*Europe's northernmost countries—
Sweden, Finland and Norway—
rank third, fourth and fifth in
Europe in terms of size. Their
combined area (429,031mi²),
however, is less than one third of
that of Canada's Northwest
Territories (1,304,900mi²).*

©DIAGRAM

BUILDINGS

Scale drawings of the plans of some famous buildings are used here to show their comparative ground areas. Our selection ranges from the Great Pyramid of Cheops, one of the world's largest ancient structures, to the Pentagon in Washington DC, which has a larger ground area than any other office building in the world.

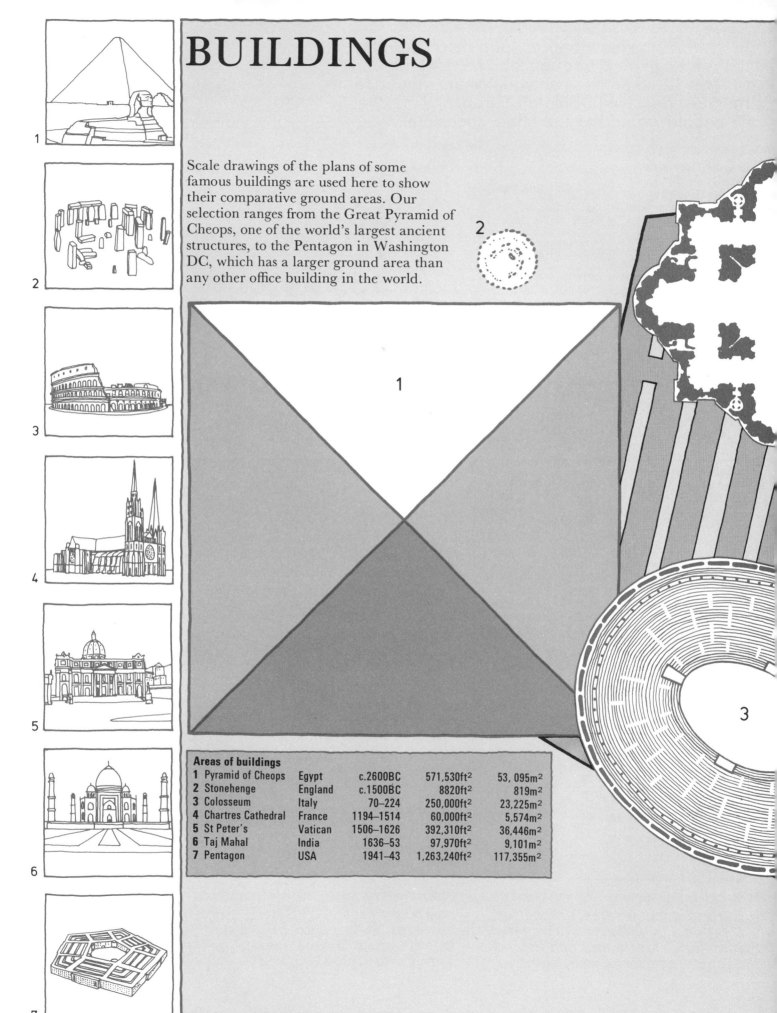

Areas of buildings				
1 Pyramid of Cheops	Egypt	c.2600BC	571,530ft^2	53,095m^2
2 Stonehenge	England	c.1500BC	8820ft^2	819m^2
3 Colosseum	Italy	70–224	250,000ft^2	23,225m^2
4 Chartres Cathedral	France	1194–1514	60,000ft^2	5,574m^2
5 St Peter's	Vatican	1506–1626	392,310ft^2	36,446m^2
6 Taj Mahal	India	1636–53	97,970ft^2	9,101m^2
7 Pentagon	USA	1941–43	1,263,240ft^2	117,355m^2

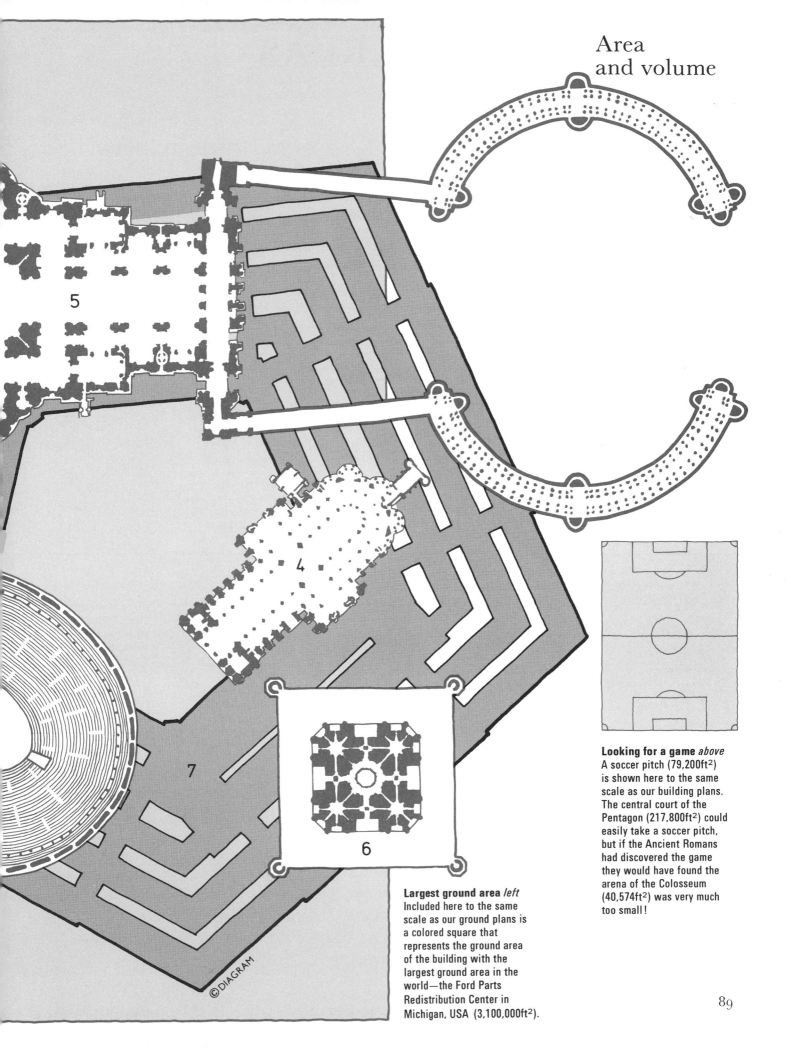

Looking for a game *above*
A soccer pitch (79,200ft^2)
is shown here to the same
scale as our building plans.
The central court of the
Pentagon (217,800ft^2) could
easily take a soccer pitch,
but if the Ancient Romans
had discovered the game
they would have found the
arena of the Colosseum
(40,574ft^2) was very much
too small!

Largest ground area *left*
Included here to the same
scale as our ground plans is
a colored square that
represents the ground area
of the building with the
largest ground area in the
world—the Ford Parts
Redistribution Center in
Michigan, USA (3,100,000ft^2).

SPORTS AREAS

The area in which a sport is played is an important element in determining the character of play. Playing area dimensions are therefore closely defined in official sports rules. In general, size is a compromise between an area in which the sport can be easily played, and an area calling for the use of extra energy or skill.

Table games *right*
A pool table is the same width as a table tennis table, but 1ft longer. This makes it 5ft² larger. An English billiards table is 2ft longer than a pool table and just over 1ft wider, making its area some 23ft² greater.

Combat sports *left*
Shown are a boxing ring, and mats for other combat sports. For wrestling the actual contest area is a circle 29½ft (9m) in diameter—76yd² (64m²). The judo contest area is of similar size—an inner square with 23½ft (8m) sides.

All to scale *left*
Shown here to the same scale are sports areas from each of the three groups of sports included on these pages.
a A table tennis table and a boxing ring.
b A judo mat and a tennis court.

Olympic track *right*
The size of an Olympic track is here drawn to the same scale as a soccer pitch. The running lanes and the land they enclose have an area of 7618yd² (6370m²). This compares with an area of 8800yd² (7300m²) for a soccer pitch.

Over 37,000 people could stand on an American football field. More than 51,000 would fit on a soccer pitch.

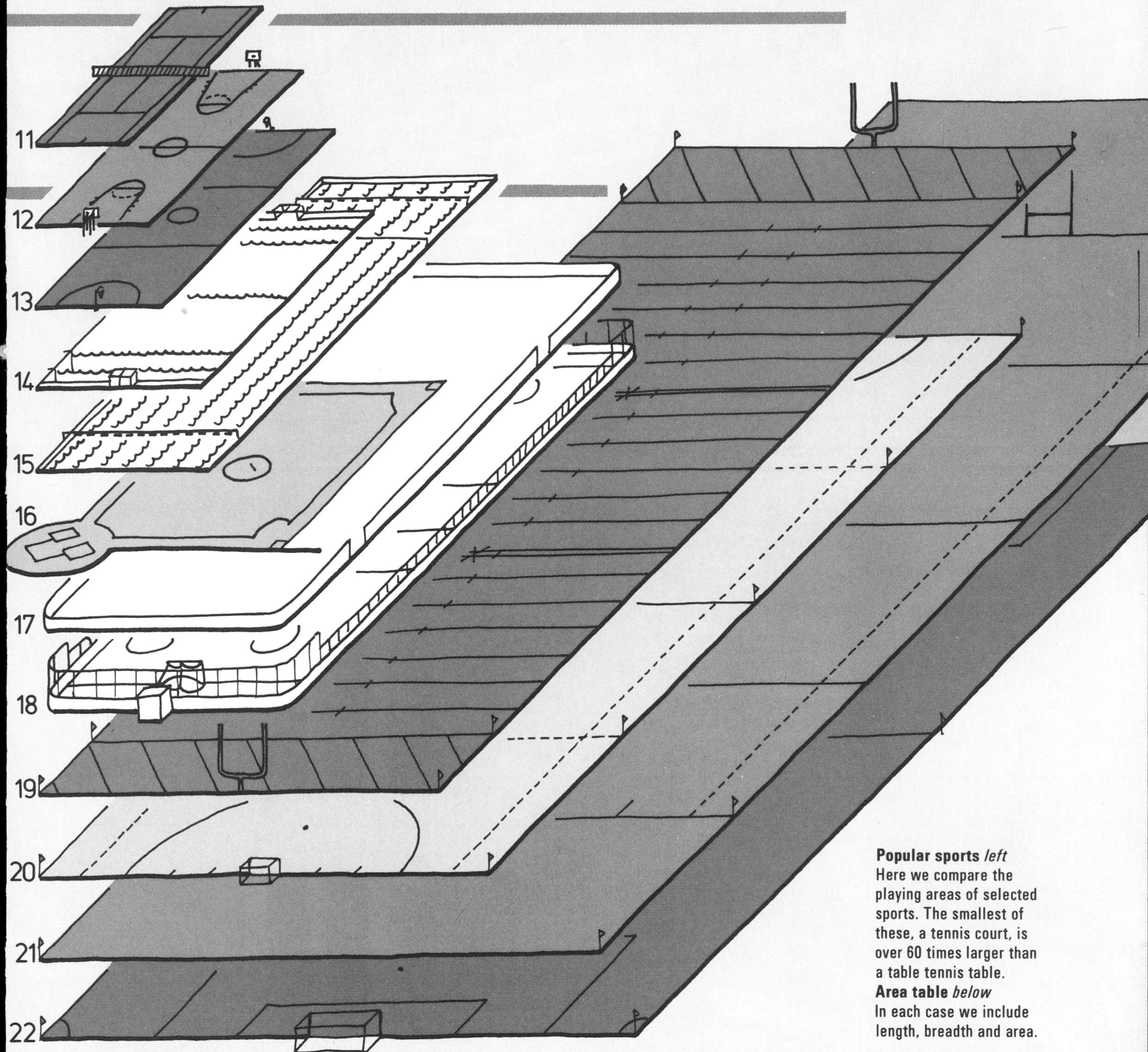

Popular sports *left*
Here we compare the
playing areas of selected
sports. The smallest of
these, a tennis court, is
over 60 times larger than
a table tennis table.
Area table *below*
In each case we include
length, breadth and area.

Sport	Dimensions	Area	Sport	Dimensions	Area
1 Table tennis	9x5ft (2.74x1.52m)	5yd² (4.16m²)	**11** Tennis	26x12yd (23.77x10.97m)	312yd² (260.75m²)
2 Pool	10x5ft (3.05x1.52m)	5.6yd² (4.63m²)	**12** Basketball	28x15yd 9in (26x14m)	427yd² (364m²)
3 Billiards, English	12x6ft 1½in (3.66x1.86m)	8.1yd² (6.80m²)	**13** Netball	33yd 1ftx16yd 2ft (30.5x15.25m)	555yd² (465.1m²)
			14 Water polo	33x22yd (30x20m)	726yd² (600m²)
			15 Swimming	55x23yd (50x21m)	1265yd² (1050m²)
4 Boxing	20x20ft (6.10x6.10m)	44yd² (37m²)	**16** Baseball	30x30yd (27.45x27.45m)	900yd² (753.5m²)
5 Karate	26x26ft (8x8m)	75yd² (64m²)	**17** Ice skating	66x33yd (60x30m)	2178yd² (1800m²)
6 Aikido	29ft 6inx29ft 6in (9x9m)	96.7yd² (81m²)	**18** Ice hockey	66yd 2ftx33yd 1ft (61x30.5m)	2222yd² (1860.5m²)
7 Kendo	36x33ft (11x10m)	132yd² (110m²)	**19** US football	120x53yd 1ft (109.8x48.8m)	6399yd² (5358.2m²)
8 Wrestling	39ft 3inx39ft 3in (12x12m)	171yd² (144m²)	**20** Field hockey	100x60yd (91.5x54.9m)	6000yd² (5063.35m²)
9 Fencing	46x6ft 6in (14x2m)	33yd² (28m²)	**21** Rugby union	160x75yd (146.3x68.62m)	1200yd² (10,039m²)
10 Judo	52ft 6inx52ft 6in (16x16m)	306yd² (256m²)	**22** Soccer	110x80yd (100x73m)	8800yd² (7300m²)

MEASURING VOLUME 1

Volume is a measure of the space occupied or contained by a three-dimensional object possessing length, breadth and height. On these two pages we compare cubic units from the US/imperial and metric systems, and use a series of cubes to show progressive increases in volume that take us from the size of a sugar cube to the volume of Earth.

Cubic measures *left*
A cubic inch (**1**) and a cubic centimeter (**2**) are drawn here to give an indication of their real size. The volume of a cube equals the area of its base (length x breadth) multiplied by its height.

Units of volume *above*
Included here, with alternative abbreviations in brackets, are units of volume obtained by "cubing" units of linear measurement from the US/imperial and metric systems. (Other units of volume are given on p. 94.)

Conversion tables *right*
To convert US/imperial volume measurements into metric, and vice versa, find the figure to be converted—US/imperial or metric—in the central column. Its equivalent will be in the appropriate column to the right or the left.

1cm^3 1000cm^3 1m^3 1000m^3 1,000,000m^3

A   B   C   D   E   F

A   B   C   D   E   F

Increasing volumes *above*
A series of cubes drawn to a constant size is here used to represent progressively larger cubes. Note that a tenfold increase in the length of a cube's sides produces a cube whose volume is 1000 times that of its predecessor. The volumes of our cubes are also used to provide fixed points on the logarithmic scale below them, and illustrated examples are used to give a clearer impression of the actual volumes involved.

Illustrated examples
A A sugar cube with 1cm sides is 1cm^3 in volume.
B A matchbox measuring 5x3.5x1.7cm has a volume of 29.75cm^3.
C A large traveling bag (80x50x25cm) is 100,000cm^3.
D A typical two-storied, four bed-roomed house with a 150m^2 ground area has a volume of roughly 1000m^3.
E The volume of St Paul's Cathedral in London is approximately 190,000m^3.
F The John F. Kennedy Space Center, Florida, boasts a Vehicle Assembly Building with a capacity of 3,666,500m^3.

The volume of a typical two-storied, four-bedroomed house (1000m^3) would fit 3666 times into the massive Vehicle Assembly Building at Florida's John F. Kennedy Space Center.

The estimated volume of Earth's oceans (1,285,600,000km^3) is over 36 times as great as the estimated volume of freshwater on Earth (35,000,000km^3).

The world's largest pyramid, the Quetzalcóatl at Cholula de Rivadabia in Mexico, had an estimated volume of 3,300,000m³ — compared with the Pyramid of Cheops, which has an estimated volume of 2,500,000m³.

in³		cm³	ft³		m³
0.0610	1	16.387	35.315	1	0.0283
0.1220	2	32.774	70.629	2	0.0566
0.1831	3	49.161	105.94	3	0.0849
0.2441	4	65.548	141.26	4	0.1133
0.3051	5	81.935	176.57	5	0.1416
0.3661	6	98.322	211.89	6	0.1699
0.4272	7	114.71	247.20	7	0.1982
0.4882	8	131.10	282.52	8	0.2265
0.5492	9	147.48	317.83	9	0.2549
0.9154	15	245.81	529.72	15	0.4248
1.5256	25	409.68	882.87	25	0.7079
2.1358	35	573.55	1236.0	35	0.9911
2.7461	45	737.42	1589.2	45	1.2743
3.3563	55	901.29	1942.3	55	1.5574
3.9665	65	1065.2	2295.5	65	1.8406
4.5768	75	1229.0	2648.6	75	2.1238
5.1870	85	1392.9	3001.7	85	2.4069
5.7973	95	1556.8	3354.9	95	2.6901

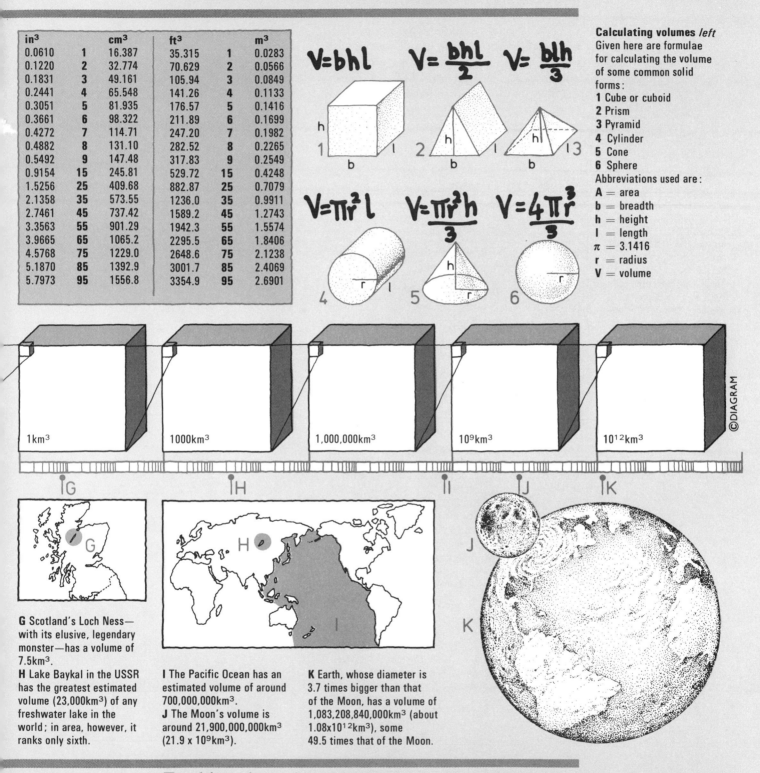

$$V = bhl$$
$$V = \frac{bhl}{2}$$
$$V = \frac{blh}{3}$$

1 2 3

$$V = \pi r^2 l$$
$$V = \frac{\pi r^2 h}{3}$$
$$V = \frac{4\pi r^3}{3}$$

4 5 6

Calculating volumes *left*
Given here are formulae for calculating the volume of some common solid forms:
1 Cube or cuboid
2 Prism
3 Pyramid
4 Cylinder
5 Cone
6 Sphere
Abbreviations used are:
A = area
b = breadth
h = height
l = length
π = 3.1416
r = radius
V = volume

©DIAGRAM

1km³   1000km³   1,000,000km³   10⁹km³   10¹²km³

G   H   I   J   K

G Scotland's Loch Ness—with its elusive, legendary monster—has a volume of 7.5km³.
H Lake Baykal in the USSR has the greatest estimated volume (23,000km³) of any freshwater lake in the world; in area, however, it ranks only sixth.

I The Pacific Ocean has an estimated volume of around 700,000,000km³.
J The Moon's volume is around 21,900,000,000km³ (21.9 x 10⁹km³).

K Earth, whose diameter is 3.7 times bigger than that of the Moon, has a volume of 1,083,208,840,000km³ (about 1.08x10¹²km³), some 49.5 times that of the Moon.

Earth's volume is equivalent to 1,083,208,840,000,000,000,000,000,000 (approximately 10²⁷) sugar lumps.

MEASURING VOLUME 2

Here we look at a further aspect of measuring volume: the use of special units for measuring the capacity of containers and thus also the quantity of substances that can be held within them. US and UK systems differ markedly in this area: the names of the units are the same, but the volumes are different in every case.

Units of capacity *right*
Listed in the table are capacity units from the US customary, UK/imperial, and metric systems of measurement. Recognized abbreviations are given in brackets. The final column gives equivalent volumes in cubic units, and serves to demonstrate the extent to which units with common names vary between the systems. Note that in the US system there are 4 fluid ounces in a gill, whereas in the UK/imperial system there are 5; otherwise multiples of units are generally similar.

US units of liquid capacity
60 minims (min)	= 1 fluid dram (fl.dr)	= 0.2256in³
8 fluid drams	= 1 fluid ounce (fl.oz)	= 1.8047in³
4 fluid ounces	= 1 gill (gi)	= 7.2187in³
4 gills	= 1 pint (pt)	= 28.875in³
2 pints	= 1 quart (qt)	= 57.750in³
4 quarts	= 1 gallon (gal)	= 231.00in³

US units of dry capacity
1 dry pint (dry pt)	= ½ dry quart (dry qt)	= 33.600in³
2 dry pints	= 1 dry quart	= 67.201in³
8 dry quarts	= 1 peck (pk)	= 537.60in³
4 pecks	= 1 bushel (bu)	= 2150.4in³

UK/imperial units of liquid and dry capacity
60 minims (min)	= 1 fluid drachm (fl.dr)	= 0.2167in³
8 fluid drachms	= 1 fluid ounce (fl.oz)	= 1.7339in³
5 fluid ounces	= 1 gill (gi)	= 8.6690in³
4 gills	= 1 pint (pt)	= 34.677in³
2 pints	= 1 quart (qt)	= 69.355in³
4 quarts	= 1 gallon (gal)	= 277.42in³
2 gallons	= 1 peck (pk)	= 554.84in³
4 pecks	= 1 bushel (bu)	= 2219.4in³
36 bushels	= 1 chaldron	= 7979,898in³

Metric units of capacity
1000 microliters or lambdas (λ)	= 1 mil (ml)	= 0.000001m³
10 milliliters or mils	= 1 centiliter (cl)	= 0.00001m³
10 centiliters	= 1 deciliter (dl)	= 0.0001m³
10 deciliters	= 1 liter (l)	= 0.001m³
1000 liters	= 1 stère (st)	= 1m³

Comparative capacities
right The diagram allows a quick comparison of capacity units: US pints (**A**), UK pints (**B**), and liters (**C**). Alongside the scales is a selection of common objects, chosen to illustrate the range of capacities involved.

a Wine glass
b Beer bottle
c Wine bottle
d Thermos (4-cup size)
e Mixing bowl
f Medium saucepan
g Earthenware casserole
h Large saucepan
i Pressure cooker
j Plastic bucket

Roll out which barrel? Among the diverse quantities for which the unit of a "barrel" has been used are: 32 gallons of herrings; 200lb of meat; 2cwt of butter; 5826 cubic inches of cranberries; and 36 gallons of beer!

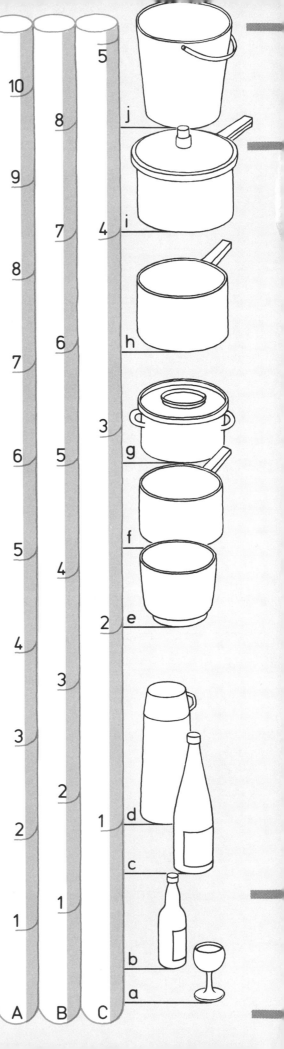

Different English versions of the Bible use three different units to express the size of the jars of water turned into wine by Jesus at the wedding at Cana (John 2:6)—two or three metretes (Wycliffe's English Bible of 1388); two or three firkins (King James Bible); twenty to thirty gallons (New English Bible). Although the units differ, the quantities are roughly similar.

US pt		UK pt	US pt		l	UK pt		l
1.2010	1	0.8327	2.1134	1	0.4732	1.7598	1	0.5683
2.4020	2	1.6653	4.2269	2	0.9463	3.5095	2	1.1365
3.6030	3	2.4980	6.3403	3	1.4195	5.2793	3	1.7048
4.8039	4	3.3306	8.4537	4	1.8926	7.0390	4	2.2730
6.0049	5	4.1633	10.567	5	2.3658	8.7988	5	2.8413
7.2059	6	4.9959	12.681	6	2.8390	10.559	6	3.4096
8.4069	7	5.8286	14.794	7	3.3121	12.318	7	3.9778
9.6079	8	6.6612	16.907	8	3.7853	14.078	8	4.5461

US gal		UK gal	US gal		l	UK gal		l
1.2009	1	0.8327	0.2642	1	3.7853	0.2200	1	4.5460
2.4019	2	1.6653	0.5284	2	7.5706	0.4400	2	9.0919
3.6029	3	2.4980	0.7925	3	11.356	0.6599	3	13.638
4.8038	4	3.3307	1.0567	4	15.141	0.8799	4	18.184
6.0047	5	4.1634	1.3209	5	18.926	1.0999	5	22.730
7.2057	6	4.9960	1.5851	6	22.712	1.3199	6	27.276
8.4066	7	5.8287	1.8492	7	26.497	1.5398	7	31.822
9.6076	8	6.6614	2.1134	8	30.282	1.7598	8	36.368
10.809	9	7.4941	2.3776	9	34.067	1.9798	9	40.914
18.014	15	12.490	3.9627	15	56.780	3.2996	15	68.189
30.024	25	20.817	6.6045	25	94.633	5.4994	25	113.65
42.033	35	29.144	9.2463	35	132.49	7.6992	35	159.11
54.043	45	37.470	11.888	45	170.34	9.8989	45	181.84
66.052	55	45.797	14.530	55	208.25	12.099	55	204.57
78.062	65	54.124	17.172	65	246.04	14.298	65	295.49
90.071	75	62.451	19.813	75	283.90	16.498	75	340.95
102.08	85	70.777	22.455	85	319.46	18.698	85	386.41
114.09	95	79.104	25.097	95	359.60	20.898	95	431.87

US fl.oz		UK fl.oz	US fl.oz		cl	UK fl.oz		cl
0.9608	1	1.0408	0.3381	1	2.9573	0.3520	1	2.8413
1.9216	2	2.0816	0.6763	2	5.9145	0.7039	2	5.6826
2.8824	3	3.1224	1.0144	3	8.8718	1.0559	3	8.5239
3.8431	4	4.1633	1.3526	4	11.829	1.4078	4	11.365
4.8039	5	5.2041	1.6907	5	14.786	1.7598	5	14.207
5.7647	6	6.2449	2.0289	6	17.744	2.1117	6	17.048
6.7255	7	7.2857	2.3670	7	20.701	2.4637	7	19.889
7.6863	8	8.3265	2.7052	8	23.658	2.8156	8	22.730
8.6471	9	9.3673	3.0433	9	26.615	3.1676	9	25.572

Conversion tables *left*
Included here are tables for converting US and UK/imperial liquid capacity measurements from one to the other, as well as tables for converting into metric units and vice versa. To use the tables, first find the figure to be converted—US, UK or metric—in the central column of the relevant table. Its equivalent can then be found in the appropriate column, either to the right or the left as indicated by the headings above.

US cookery measures	Metric equivalents
1 teaspoon	0.5cl
1 tablespoon (= 3 teaspoons)	1.5cl
1 cup (= 16 tablespoons = ½ US pint)	23.7cl
UK cookery measures	
1 teaspoon	0.6cl
1 dessertspoon (= 2 teaspoons)	1.2cl
1 tablespoon (= 3 teaspoons)	1.8cl
1 cup (= 16 tablespoons = ½ UK pint)	28.4cl

Recipe for success
The table *above* gives the centiliter equivalents of the standard measuring spoons and cups commonly used by cooks in the USA and UK. Note that the standard measuring cup is a "breakfast" cup rather than a teacup, which is half its size. As the diagram *right* further illustrates, US cooks' measures are roughly four-fifths as big as UK measures.
a UK tablespoon
b US tablespoon
c UK dessertspoon
d UK teaspoon
e US teaspoon

1·5 cl
1·0 cl
0·5 cl
a
b
c
d
e

For a bubbly bath *right*
The standard champagne bottle holds 80cl (5–10cl more than standard bottles for other wines). Champagne is also sold in larger bottles, of which the names and capacities (given as multiples of the standard 80cl bottle) are given here.

a Magnum (2)
b Jeroboam (4)
c Rehoboam (6)
d Methuselah (8)
e Salmanazar (12)
f Balthazar (16)
g Nebuchadnezzar (20)

a
b
c
d
e
f
g

If Scotland's Loch Ness were to be emptied in search of the monster, it would take 1650 billion UK one-gallon buckets, or 1981 billion US one-gallon buckets.

OIL AND WATER

Man's largest oil tankers and storage
tanks are tiny compared with the volumes
of water contained in nature. Even the
man-made lake with the greatest volume,
at Owen Falls, would fit 113 times into the
world's most voluminous natural freshwater
lake (Baykal 5520mi³), and could be filled
in 19½ weeks by the flow over Boyoma Falls.

A	Pierre Guillaumat	678,000m³	3,307,317
B	Bellamya	677,362m³	3,304,205
C	Batillus	677,362m³	3,304,205
D	Globtik London	578,235m³	2,820,658
E	Globtik Tokyo	573,361m³	2,796,883
F	Nissei Maru	573,345m³	2,796,805
G	Esso Mediterranean	553,054m³	2,697,824
H	Berge Emperor	513,680m³	2,505,756
I	Berge Empress	513,680m³	2,505,756
J	Al Rekkah	512,920m³	2,502,049

Oil in the tanks
The table *above* and the
diagram *left* show the
oil-carrying capacity of the
world's 10 largest
supertankers. In the table,
the ships' capacities are
expressed first in cubic
meters (in accordance with
international practice in
this area), and also in terms
of standard international oil
barrels (each with a
capacity of 205 liters).

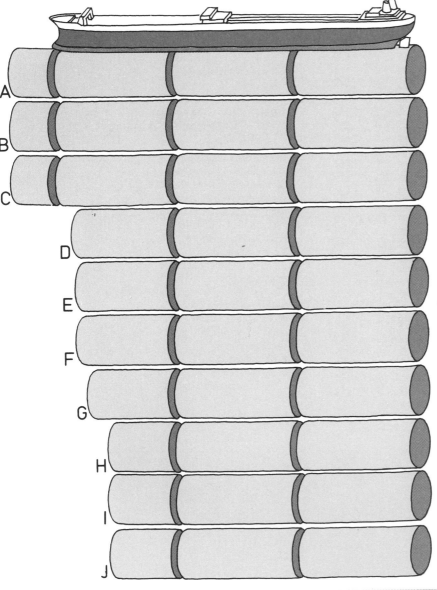

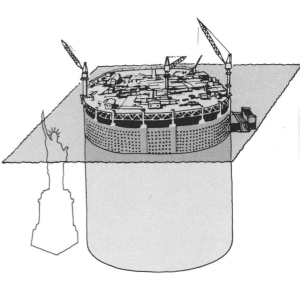

Giant oil tank *above*
The world's largest oil
storage tank—in the North
Sea's Ekofisk field—is shown
here to the same scale as the
Statue of Liberty. Note,
however, that the overall
volume of the Ekofisk tank
is roughly five times as
great as the tank's oil
capacity. When full, the
tank holds some 160,000m³
of oil, equal to 780,466
standard international 205l
oil barrels, or almost
one-quarter of the capacity
of the world's largest oil
tanker, France's *Pierre
Guillaumat* (678,000m³).

**The capacity of the world's
largest oil tanker, the *Pierre
Guillaumat*, is equivalent to the
tanks of nearly 10 million
average-sized US automobiles.**

If the 48.5 cubic miles of water in the man-made lake at Owen Falls, Uganda, were to be put in a regularly shaped tank, the tank would be a cube with edges of 3.65 miles.

a	Boyoma (Stanley)	Zaire	600,000ft³/s	0.47s
b	Guaíra	Brazil/Paraguay	470,000ft³/s	0.60s
c	Khône	Laos	410,000ft³/s	0.69s
d	Niagara	Canada/USA	212,000ft³/s	1.33s
e	Paulo Afonso	Brazil	100,000ft³/s	2.82s
f	Urubupungá	Brazil	97,000ft³/s	2.91s
g	Cataratas del Iguazú	Brazil/Argentina	61,660ft³/s	4.57s
h	Patos-Maribondo	Brazil	53,000ft³/s	5.32s
i	Victoria	Zimbabwe-Rhodesia	38,430ft³/s	7.34s
j	Churchill (Grand)	Canada	35,000ft³/s	8.06s

©DIAGRAM

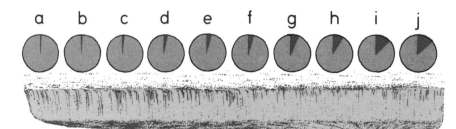

a b c d e f g h i j

Over the falls *above*
The table shows the world's top 10 waterfalls in terms of volume. The first column of figures gives their mean annual water flow, expressed in cubic feet per second, or cusecs (ft³/s). The other column of figures, and the accompanying diagram, show the amount of time, in seconds, that would be needed for each of these waterfalls to fill the dome of London's St Paul's Cathedral (282,000ft³).

1	Owen Falls	Victoria Nile, Uganda	48.5mi³
2	Bratsk	Angara, USSR	40.6mi³
3	Aswan High	Nile, Egypt	39.4mi³
4	Kariba	Zambesi, Zambia/Rhodesia	38.4mi³
5	Akosombo	Volta, Ghana	35.5mi³
6	Daniel Johnson	Manicouagan, Canada	34.1mi³
7	Krasnoyarsk	Yenisey, USSR	17.6mi³
8	WAC Bennett	Peace, Canada	16.8mi³
9	Zeya	Zeya, USSR	16.4mi³
10	Wadi Tharthar	Tigris, Iraq	16.0mi³

Behind the dams
The table *above* gives the names of the world's greatest dams, the rivers that they dam, the countries in which they are situated and the volume of water in the lakes behind them. The diagram *left* provides a visual comparison of these volumes. Greatest among them is the volume of water in the lake behind Uganda's Owen Falls— some 48.5mi³, a volume sufficient to fill more than 1,762,000,000,000 average bathtubs!

1 2 3 4 5 6 7 8 9 10

Here we see at last the head of the 22ft-long giant earthworm whose tail is on page 66.

A fast-running domestic tap would take 308 days to fill the dome of London's St Paul's Cathedral.

THE UNIVERSE

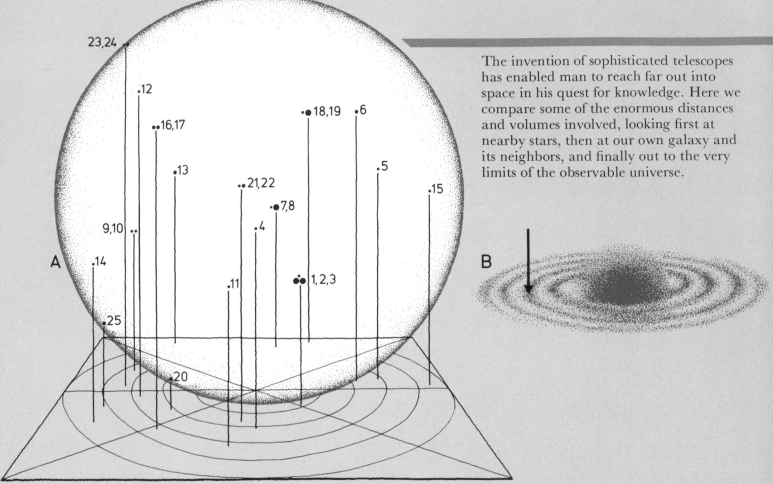

23,24
.12
.16,17
.13
.18,19 .6
.5
9,10
.21,22
.7,8
.4
.15
A
.14
.11
.25
.1,2,3
.20

The invention of sophisticated telescopes has enabled man to reach far out into space in his quest for knowledge. Here we compare some of the enormous distances and volumes involved, looking first at nearby stars, then at our own galaxy and its neighbors, and finally out to the very limits of the observable universe.

B

Light years away *below right* Here we use spheres of constant size to represent progressively larger spheres. Each new sphere has a radius (**r**) 10 times bigger than that of its predecessor, giving it a volume (**V**) that is 1000 times as great. Dimensions are expressed in light years (ly) and cubic light years (ly³). Drawn within four of the spheres are colored spheres (**A,B,C,D**) indicating the comparative volumes of the various space features shown in more detail *above*.
A Sphere with radius of 12 ly, containing our Sun's stellar neighbors.
B Sphere representing the volume of our galaxy.
C Sphere with radius of 2,500,000 ly, containing our neighboring galaxies.
D Sphere representing the theoretical edge of the universe.

A) Stellar neighbors *above* This sphere, with its center at our Sun, has a radius of 12 ly and a volume of 7238 ly³. Stellar neighbors situated within these limits are listed *right* and located both within the sphere and on the plan view below it.

1 Proxima Centauri	**11** Ross 154	**21** Sigma 2398 A
2 Alpha Centauri	**12** Ross 248	**22** Sigma 2398 B
3 Beta Centauri	**13** Epsilon Eridani	**23** Groombridge 34A
4 Barnard's star	**14** Lutyens 789-6	**24** Groombridge 34B
5 Wolf 359	**15** Ross 128	**25** Lacertae 9352
6 LAL 21185	**16** 61 Cygni A	
7 Sirius A	**17** 61 Cygni B	
8 Sirius B	**18** Procyon A	
9 UV Ceti A	**19** Procyon B	
10 UV Ceti B	**20** Epsilon Indi	

B) Volume of our galaxy *above* Here we show our galaxy, with its well-defined nucleus and spiral arms. The Sun's approximate position is indicated by an arrow. Our galaxy's volume (mass divided by density) is about 3.7×10^{13} ly³, represented *bottom* by sphere **B**.

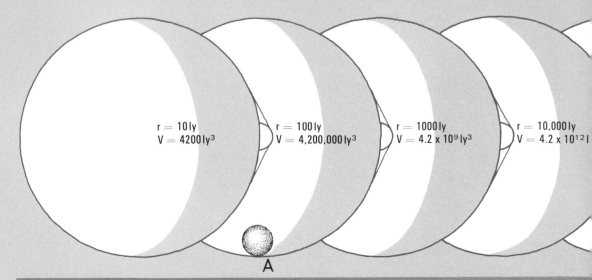

r = 10 ly
V = 4200 ly³

r = 100 ly
V = 4,200,000 ly³

r = 1000 ly
V = 4.2×10^9 ly³

r = 10,000 ly
V = 4.2×10^{12} ly³

A

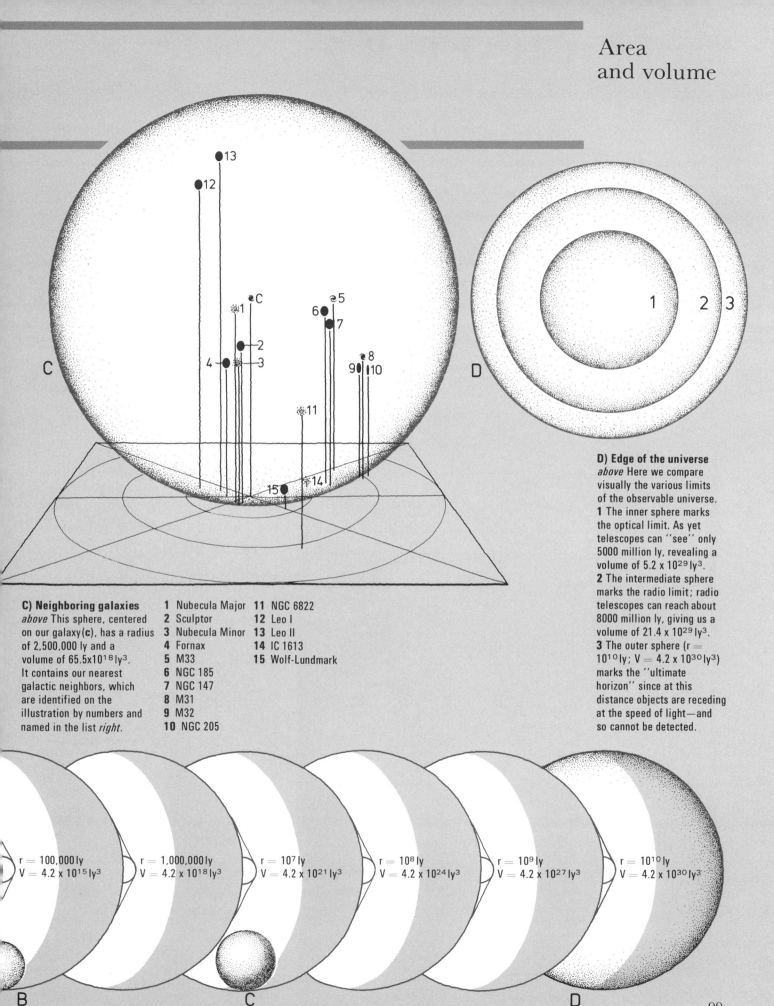

C) Neighboring galaxies
above This sphere, centered on our galaxy (**c**), has a radius of 2,500,000 ly and a volume of 65.5x10^{18} ly^3. It contains our nearest galactic neighbors, which are identified on the illustration by numbers and named in the list *right*.

1	Nubecula Major	**11**	NGC 6822
2	Sculptor	**12**	Leo I
3	Nubecula Minor	**13**	Leo II
4	Fornax	**14**	IC 1613
5	M33	**15**	Wolf-Lundmark
6	NGC 185		
7	NGC 147		
8	M31		
9	M32		
10	NGC 205		

D) Edge of the universe
above Here we compare visually the various limits of the observable universe.
1 The inner sphere marks the optical limit. As yet telescopes can "see" only 5000 million ly, revealing a volume of 5.2 x 10^{29} ly^3.
2 The intermediate sphere marks the radio limit; radio telescopes can reach about 8000 million ly, giving us a volume of 21.4 x 10^{29} ly^3.
3 The outer sphere (r = 10^{10} ly; V = 4.2 x 10^{30} ly^3) marks the "ultimate horizon" since at this distance objects are receding at the speed of light—and so cannot be detected.

r = 100,000 ly
V = 4.2 x 10^{15} ly^3

r = 1,000,000 ly
V = 4.2 x 10^{18} ly^3

r = 10^7 ly
V = 4.2 x 10^{21} ly^3

r = 10^8 ly
V = 4.2 x 10^{24} ly^3

r = 10^9 ly
V = 4.2 x 10^{27} ly^3

r = 10^{10} ly
V = 4.2 x 10^{30} ly^3

B

C

D

©DIAGRAM

CHAPTER 5

Drawings of Japanese Sumo wrestlers from the book *Sumo Nyakunenshi* (published by Kodansha of Tokyo). The minimum weight for a Sumo wrestler is 350 lb—more than twice the weight of an average man. Perhaps the heaviest *sumatori* of all was the 430 lb Dewayatake.

MASS, WEIGHT AND DENSITY

This photograph taken during the 1969 Apollo 11 mission shows Astronaut Edwin Aldrin walking on the surface of the Moon, where gravity—and hence a person's weight—is only about one sixth of that on Earth (NASA, Washington DC).

MEASURING MASS, WEIGHT AND DENSITY

Mass measures the amount of matter, or substance, in an object. Density is the mass per unit volume, which varies with temperature and pressure. Weight measures the pull of gravity on an object and is directly related to the object's mass. For most purposes mass and weight are the same, as the gravitational force on which weight depends is for most of us always that of Earth (although even this varies slightly with latitude and altitude, see p.112). Systems of weights use multiples and divisions of the weight of commonly agreed objects. Base weights range from a sack of grain, popular among agricultural societies, to the metric system's standard kilogram—a cylinder of platinum-iridium alloy.

Weight and gravity *left*
Weight measures the force of attraction between an object and (for us) Earth. The stronger this force, the heavier the object. It is measured most simply by a spring scale marked in commonly agreed units of weight.

Systems of weights *right*
Listed are units for measuring mass and weight. Standard abbreviations are given in brackets. "Short" or "net" hundredweights and tons are fuller names for these US units, "long" or "gross" hundredweights and tons for the UK units.

Conversion tables *right*
To convert US/imperial weight measurements into metric, and vice versa, find the figure to be converted—US/imperial or metric—in the central column. Its equivalent will be in the appropriate column to the right or to the left.

Tons and tons and tonnes *far right* These conversion tables enable us to convert short tons to tonnes, and long tons to tonnes. Short tons can be converted to long tons (and vice versa) by first converting into tonnes and then to short or to long tons.

US units of mass and weight (avoirdupois)

16 drams (dr)	= 1 ounce (oz)	
16 ounces	= 1 pound (lb)	
100 pounds	= 1 hundredweight (cwt)	
5 hundredweights	= 1 quarter (qr)	= 500 lb
4 quarters	= 1 ton (tn)	= 2000 lb

UK/imperial units of mass and weight (avoirdupois)

16 drachms (dr)	= 1 ounce (oz)	
16 ounces	= 1 pound (lb)	
14 pounds	= 1 stone (st)	
2 stones	= 1 quarter (qr)	= 28 lb
4 quarters	= 1 hundredweight (cwt)	= 112 lb
20 hundredweights	= 1 ton (tn)	= 2240 lb

Metric units of mass and weight

1000 milligrams (mg)	= 1 gram (g)
1000 grams	= 1 kilogram (kg)
1000 kilograms	= 1 tonne (t)

Troy (jewelers') units of weight

24 grains (gr)	= 1 pennyweight (dwt)	
20 pennyweights	= 1 ounce (oz.t)	= 1.097 oz
12 ounces	= 1 pound (lb.t)	= 0.823 lb

Apothecaries' units of weight

20 grains (gr)	= 1 scruple (s.ap)	
3 scruples	= 1 dram (dr.ap)	
8 drams	= 1 ounce (oz.ap)	= 1.097 oz
12 ounces	= 1 pound (lb.ap)	= 0.823 lb

oz		g	lb		kg
0.0353	1	28.350	2.2046	1	0.4536
0.0705	2	56.699	4.4092	2	0.9072
0.1058	3	85.049	6.6139	3	1.3608
0.1411	4	113.40	8.8185	4	1.8144
0.1764	5	141.75	11.023	5	2.2680
0.2116	6	170.10	13.228	6	2.7216
0.2469	7	198.45	15.432	7	3.1752
0.2822	8	226.80	17.637	8	3.6287
0.3175	9	255.15	19.842	9	4.0823
0.5291	15	425.24	33.069	15	6.8039
0.8819	25	708.74	55.116	25	11.340
1.2346	35	992.23	77.162	35	15.876
1.5873	45	1275.7	99.208	45	20.412
1.9401	55	1559.2	121.25	55	24.948
2.2928	65	1842.7	143.30	65	29.484
2.6456	75	2126.2	165.35	75	34.019
2.9983	85	2409.7	187.39	85	38.555
3.3510	95	2693.2	209.44	95	43.091

short tons		tonnes	long tons		tonnes
1.1023	1	0.9072	0.9842	1	1.0161
2.2046	2	1.8144	1.9684	2	2.0321
3.3069	3	2.7216	2.9526	3	3.0482
4.4092	4	3.6287	3.9368	4	4.0642
5.5116	5	4.5359	4.9210	5	5.0803
6.6139	6	5.4431	5.9052	6	6.0963
7.7162	7	6.3503	6.8895	7	7.1124
8.8185	8	7.2575	7.8737	8	8.1284
9.9208	9	8.1647	8.8579	9	9.1445
16.535	15	13.607	14.763	15	15.241
27.558	25	22.680	24.605	25	25.401
38.581	35	31.751	34.447	35	35.562
49.604	45	40.823	44.289	45	45.722
60.627	55	49.895	54.131	55	55.883
71.650	65	58.967	63.973	65	66.043
82.673	75	68.039	73.816	75	76.204
93.696	85	77.111	83.658	85	86.364
104.72	95	86.182	93.500	95	96.525

Use of the term carat can lead to confusion when describing an engagement ring. When applied to precious stones, such as a diamond, a carat is a unit of weight equal to 200 milligrams. When applied to gold, however, a carat (more properly spelled karat in this case) is a measure of purity not of weight, being equal to a 1/24th part of pure gold in an alloy.

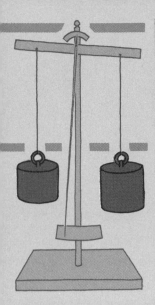

Mass and matter *above*
Mass measures the amount of matter in an object, and is measured by comparing an object with others of known mass on a balance. Mass is expressed in the same units as weight (lb, kg, etc), but unlike weight it does not depend on gravity.

Determining density
The density of an object is its mass divided by its volume. The greater the density of a substance, then the more concentrated is its matter, the more mass it will have per unit volume, and, volume for volume, the heavier it will seem.

		lb/ft³	g/cm³
1	Platinum	1338.48	21.45
2	Gold	1203.70	19.29
3	Lead	709.49	11.37
4	Silver	651.46	10.44
5	Granite	164.74	2.64
6	Concrete	134.78*	2.16*
7	Sugar	100.00*	1.61*
8	Coal	84.50*	1.35*
9	Boxwood	64.90*	1.04*
10	Milk	64.27	1.03
11	Water	62.40	1
12	Rubber	58.03	0.93
13	Petroleum	54.80	0.88
14	Alcohol	49.30	0.79
15	Beechwood	44.93*	0.72*
16	Charcoal	21.53*	0.345*
17	Air (62°F)	0.08	0.00128

* Mean density

Density, mass and volume
The table *left* lists the densities of selected substances. The diagram *above* shows, real size, the volumes that 4g (about the weight of an airmail envelope and one sheet of paper) of these substances would occupy.

The densest elements
below This table shows the 10 densest elements at 20°C. The order depends on temperature because different substances expand and contract at different rates in response to the addition or subtraction of heat.

1	Osmium	1409.62 lb/ft³	22.59g/cm³
2	Iridium	1407.74 lb/ft³	22.56g/cm³
3	Platinum	1338.48 lb/ft³	21.45g/cm³
4	Rhenium	1311.02 lb/ft³	21.01g/cm³
5	Gold	1203.70 lb/ft³	19.29g/cm³
6	Tungsten	1201.82 lb/ft³	19.26g/cm³
7	Uranium	1188.72 lb/ft³	19.05g/cm³
8	Tantalum	1040.21 lb/ft³	16.67g/cm³
9	Protactinium	961.58 lb/ft³	15.41g/cm³
10	Mercury	845.52 lb/ft³	13.55g/cm³

Relative density *right*
Comparisons of density can be made with reference to the "specific gravity" of different substances. The specific gravity (sg) of water is 1; the sg of other substances is the ratio, at a certain temperature (usually 60°F), of the weight of the substance to the weight of the same volume of water. The weight of the same volume of water is the difference between the weight of the substance when weighed in air (**a**) or in water (**b**). More often, sg is measured with a special instrument—a hydrometer.

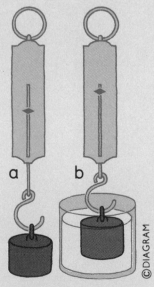

©DIAGRAM

A table top measuring 6ft by 2ft by 1in thick would weigh 32lb if made of yellow pine. If made of marble it would weigh over five times as much (165lb).

COMPARATIVE MASSES

On these two pages we look at the whole range of masses, from the mass of a neutrino, the smallest known mass at 5×10^{-34} kg, to the largest, which is the estimated mass of the universe, at 10^{51} kg. Masses within the living world range from that of a virus at 10^{-21} kg, to that of a blue whale at 1.38×10^5 kg.

Universal scale of mass
below The logarithmic scale across these pages shows the whole range of known or estimated masses, allowing us to compare the masses of things as diverse as a subatomic particle, a hummingbird, a Chieftain tank, and the Moon.

Enlarged sections of the main scale have made it possible to plot a greater number of examples, and so allow more detailed comparisons of very tiny masses, of the masses of living and man-made things, and of the huge masses to be found in space.

Mass in the living world
below Plotted on this enlarged section of the scale are examples showing the wide range of masses in the living world.
8 Virus 10^{-21} kg.
9 Bacteria 10^{-13} to 10^{-14} kg.
10 Parasitic wasp, the smallest insect, 5×10^{-9} kg.
11 House spider 10^{-4} kg.
12 Helena's hummingbird 2×10^{-3} kg.
13 Chicken 3.15 kg.
14 Average man 7.29×10 kg.
15 Polar bear 3.22×10^2 kg.
16 African elephant 6.3×10^3 kg.
17 Blue whale, the most massive animal, 1.38×10^5 kg.

8

17

Universal scale of mass

| 10^{-20} | 10^{-15} | 10^{-10} | 10^{-5} | 1 |

| 10^{-30} | 10^{-25} | | 1 | 10 |

Minute masses *above*
Plotted on this enlarged section are the masses of subatomic particles, atoms and molecules.
1 Neutrino, a stable, electrically neutral particle, 5×10^{-34} kg.
2 Electron 9.1096×10^{-31} kg.
3 Proton 1.6726×10^{-27} kg.
4 Neutron, a long-lived, neutral particle, 1.6748×10^{-27} kg.
5 Carbon 12 atom 1.9924×10^{-26} kg.
6 Molecule of water (H_2O) 2.99×10^{-26} kg.
7 Molecule of a complex biochemical compound 10^{-22} kg.

5

18   19   20   21   22   23

The mass of a virus is to the mass of an average man as the mass of an average man is to the mass of Earth.

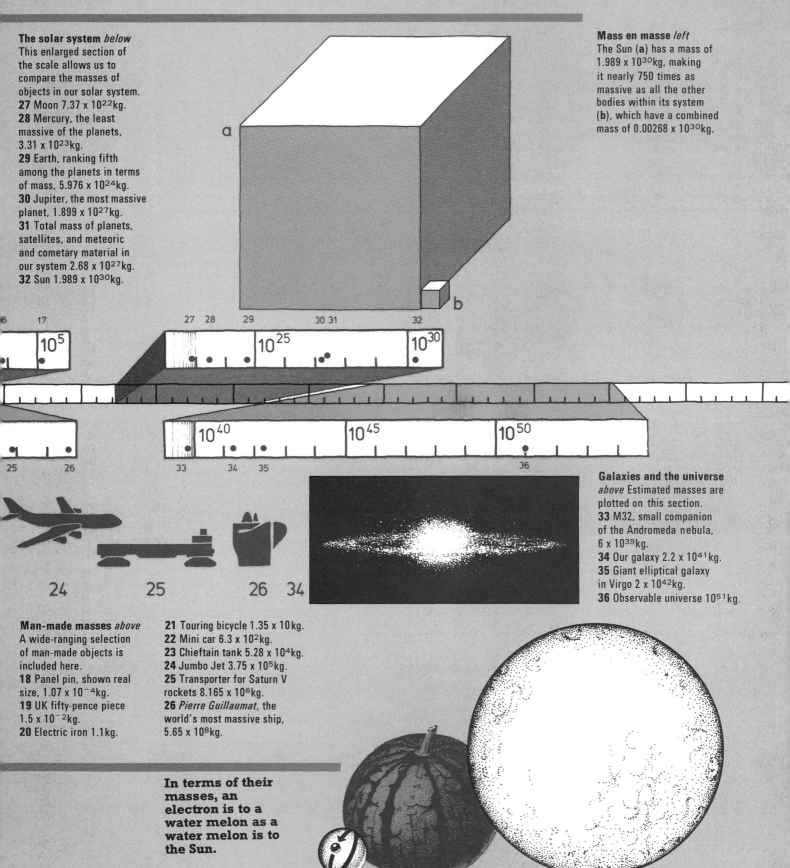

Among subatomic particles, the mass of a neutron, at 1.6748×10^{-27} kg, is over three million times greater than the mass of a neutrino, at 5×10^{-34} kg.

The solar system *below*
This enlarged section of the scale allows us to compare the masses of objects in our solar system.
27 Moon 7.37×10^{22}kg.
28 Mercury, the least massive of the planets, 3.31×10^{23}kg.
29 Earth, ranking fifth among the planets in terms of mass, 5.976×10^{24}kg.
30 Jupiter, the most massive planet, 1.899×10^{27}kg.
31 Total mass of planets, satellites, and meteoric and cometary material in our system 2.68×10^{27}kg.
32 Sun 1.989×10^{30}kg.

Mass en masse *left*
The Sun (**a**) has a mass of 1.989×10^{30}kg, making it nearly 750 times as massive as all the other bodies within its system (**b**), which have a combined mass of 0.00268×10^{30}kg.

Galaxies and the universe
above Estimated masses are plotted on this section.
33 M32, small companion of the Andromeda nebula, 6×10^{39}kg.
34 Our galaxy 2.2×10^{41}kg.
35 Giant elliptical galaxy in Virgo 2×10^{42}kg.
36 Observable universe 10^{51}kg.

Man-made masses *above*
A wide-ranging selection of man-made objects is included here.
18 Panel pin, shown real size, 1.07×10^{-4}kg.
19 UK fifty-pence piece 1.5×10^{-2}kg.
20 Electric iron 1.1kg.

21 Touring bicycle 1.35×10kg.
22 Mini car 6.3×10^{2}kg.
23 Chieftain tank 5.28×10^{4}kg.
24 Jumbo Jet 3.75×10^{5}kg.
25 Transporter for Saturn V rockets 8.165×10^{6}kg.
26 *Pierre Guillaumat*, the world's most massive ship, 5.65×10^{8}kg.

In terms of their masses, an electron is to a water melon as a water melon is to the Sun.

MATTER

All matter consists of atoms, which combine to form "elements" and "compounds." Here we use the "periodic table" as a means of comparing elements. Substances as different as hydrogen, calcium and gold are all elements; each has distinctive chemical and physical properties and cannot be split chemically into simpler form.

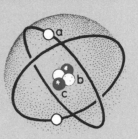

Atomic structure *left*
Electrons (**a**) orbit a nucleus of protons (**b**) and neutrons (**c**). Electrons have a negative charge and protons a positive charge; neutrons are neutral. There are always the same number of electrons as protons; the number of neutrons varies.

How many elements?
below When the Russian scientist Mendeleev drew up the first periodic table in 1869, only 65 elements were known. In the modern periodic table there are 104 known elements. 88 of these, from hydrogen to uranium, occur naturally, 16, including elements 43, 61, 85 and 87, are man-made or occur instantaneously during radioactive decay. Further elements have been predicted, and names already prepared for them.

The pattern of the table
below The table groups the elements into seven lines, or periods. In each horizontal line there is a repetition of chemical properties: as we read from left to right the elements of each line become less metallic.
The elements of each vertical group in the main table also have similar chemical properties.

The "lanthanides" or rare earths (numbers 57–71) and the "actinides" (numbers 89–103) form separate groups because, despite their atomic numbers, their properties are so similar that they fit the space of only one element in the main table.

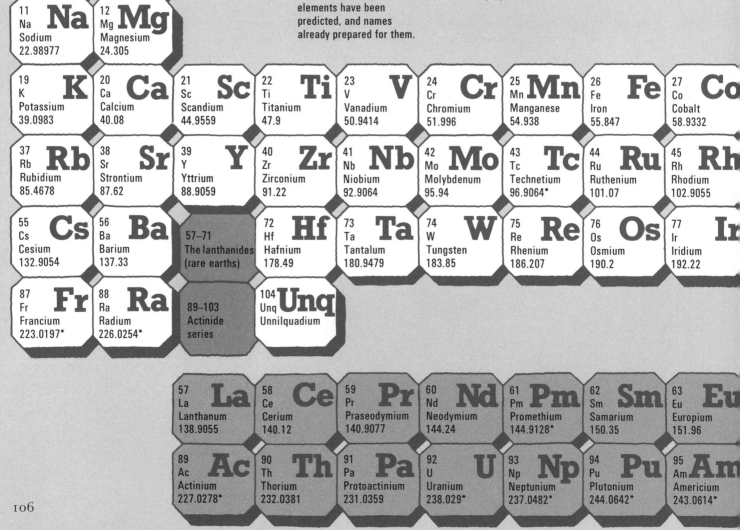

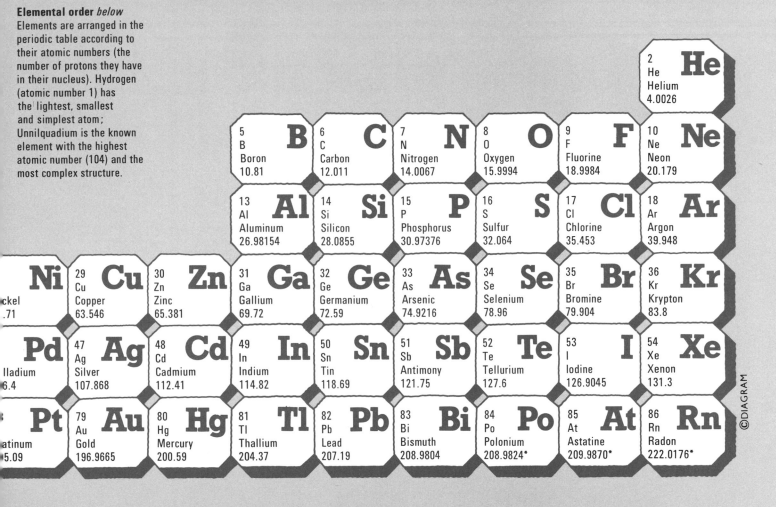

Explanation of symbols and figures *left*
a Atomic number
b Letter symbol for each element
c Element's name

d Atomic mass number (or atomic weight) in international amu's
* Atomic weight of the isotope with the longest known half-life

a → 99
b → Es
c → Einsteinium
d → 254.0880*

Es

Atomic mass units (amu)	Derivation	Equivalent in kg
1 amu (international) =	One-twelfth the mass of carbon 12, the principle isotope of carbon. =	1.66033×10^{-27} kg
1 amu (physical) =	One-sixteenth the mass of oxygen 16 (99.8% of all oxygen found on Earth). =	1.65981×10^{-27} kg
1 amu (chemical) =	One-sixteenth of the average mass of a mixture of three oxygen isotopes. =	1.66026×10^{-27} kg

Atomic mass units *left*
An atom's mass (or weight) is measured in multiples of an atomic mass unit (amu). This table shows the three types of amu. Note that the amu differs from the "atomic unit of mass," which is the electron's rest mass: 9.1084×10^{-31} kg.

Elemental order *below*
Elements are arranged in the periodic table according to their atomic numbers (the number of protons they have in their nucleus). Hydrogen (atomic number 1) has the lightest, smallest and simplest atom; Unnilquadium is the known element with the highest atomic number (104) and the most complex structure.

2 He Helium 4.0026 **He**

5 B Boron 10.81 **B**
6 C Carbon 12.011 **C**
7 N Nitrogen 14.0067 **N**
8 O Oxygen 15.9994 **O**
9 F Fluorine 18.9984 **F**
10 Ne Neon 20.179 **Ne**

13 Al Aluminum 26.98154 **Al**
14 Si Silicon 28.0855 **Si**
15 P Phosphorus 30.97376 **P**
16 S Sulfur 32.064 **S**
17 Cl Chlorine 35.453 **Cl**
18 Ar Argon 39.948 **Ar**

Ni ckel .71
29 Cu Copper 63.546 **Cu**
30 Zn Zinc 65.381 **Zn**
31 Ga Gallium 69.72 **Ga**
32 Ge Germanium 72.59 **Ge**
33 As Arsenic 74.9216 **As**
34 Se Selenium 78.96 **Se**
35 Br Bromine 79.904 **Br**
36 Kr Krypton 83.8 **Kr**

Pd lladium 6.4
47 Ag Silver 107.868 **Ag**
48 Cd Cadmium 112.41 **Cd**
49 In Indium 114.82 **In**
50 Sn Tin 118.69 **Sn**
51 Sb Antimony 121.75 **Sb**
52 Te Tellurium 127.6 **Te**
53 I Iodine 126.9045 **I**
54 Xe Xenon 131.3 **Xe**

Pt atinum 5.09
79 Au Gold 196.9665 **Au**
80 Hg Mercury 200.59 **Hg**
81 Tl Thallium 204.37 **Tl**
82 Pb Lead 207.19 **Pb**
83 Bi Bismuth 208.9804 **Bi**
84 Po Polonium 208.9824* **Po**
85 At Astatine 209.9870* **At**
86 Rn Radon 222.0176* **Rn**

©DIAGRAM

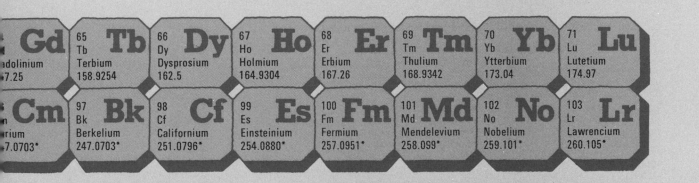

Gd adolinium 7.25
65 Tb Terbium 158.9254 **Tb**
66 Dy Dysprosium 162.5 **Dy**
67 Ho Holmium 164.9304 **Ho**
68 Er Erbium 167.26 **Er**
69 Tm Thulium 168.9342 **Tm**
70 Yb Ytterbium 173.04 **Yb**
71 Lu Lutetium 174.97 **Lu**

Cm rium 7.0703*
97 Bk Berkelium 247.0703* **Bk**
98 Cf Californium 251.0796* **Cf**
99 Es Einsteinium 254.0880* **Es**
100 Fm Fermium 257.0951* **Fm**
101 Md Mendelevium 258.099* **Md**
102 No Nobelium 259.101* **No**
103 Lr Lawrencium 260.105* **Lr**

HARDNESS

Solids vary in their degree of hardness or softness. Their degree of hardness depends on their microscopic internal structure—the arrangement of the atoms from which they are made. Many solids are in the form of crystals, whose macroscopic, flat-sided, geometric shapes indicate the microscopic arrangement of their atoms.

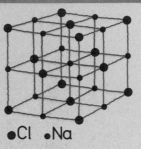

● Cl ● Na

Building with atoms *left*
The regular shape of a crystal is an indication of its atomic structure. Salt crystals are cube-shaped because they are made up of cubic arrangements of alternating sodium (Na) and chlorine (Cl) atoms.

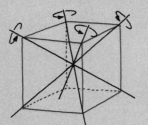
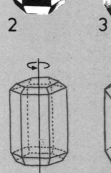
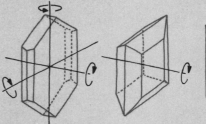

1   2   3   4   5   6   7

Crystal symmetry
Illustrated *above* and included in the table *right* are examples of seven basic types or "systems," of crystal. A crystal's system depends on the symmetry of its structure, which can be determined, as here, from its "axes of symmetry." On the schematic diagram beneath each crystal we show the minimum number of axes of symmetry needed to classify a crystal within a system. These axes may be 2-, 3-, 4- or 6-fold depending on how many times the crystal presents an identical aspect when rotated through 360°.

System	Minimum symmetry	Example	Composition	Uses
1 Cubic or isometric	Four 3-fold axes	Diamond	Carbon	Drills, abrasives, cutting tools, gems
2 Hexagonal	One 6-fold axis	Beryl (emerald)	Beryllium aluminum silicate	Gems
3 Trigonal	One 3-fold axis	Tourmaline	Aluminum borosilicate	Electronics, gems, pressure gauges
4 Tetragonal	One 4-fold axis	Cassiterite	Tin dioxide or tinstone	Source of tin
5 Orthorhombic	Three 2-fold axes	Barite	Barium sulfate	Source of barium compounds
6 Monoclinic	One 2-fold axis	Gypsum	Hydrated calcium sulfate	Plaster of Paris, cement, paint
7 Triclinic	None	Albite	Sodium aluminum silicate	Glass, ceramics

In the words of the song, "Diamonds are a girl's best friend." Certainly they are long lasting, being the hardest known naturally occurring substance. Among other precious stones, sapphires and rubies both rank second in terms of hardness, and emeralds third.

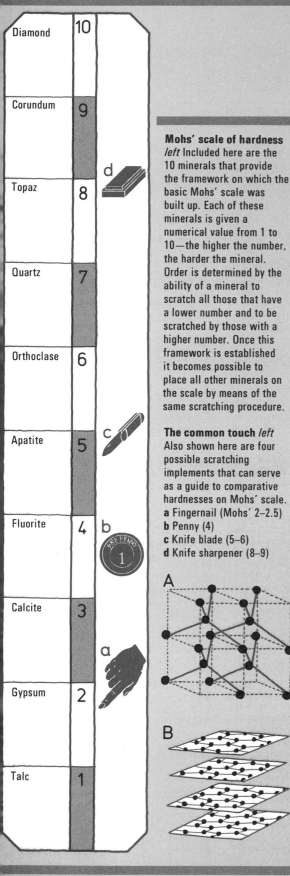

Mineral	Mohs
Diamond	10
Corundum	9
Topaz	8
Quartz	7
Orthoclase	6
Apatite	5
Fluorite	4
Calcite	3
Gypsum	2
Talc	1

Mohs' scale of hardness
left Included here are the 10 minerals that provide the framework on which the basic Mohs' scale was built up. Each of these minerals is given a numerical value from 1 to 10—the higher the number, the harder the mineral. Order is determined by the ability of a mineral to scratch all those that have a lower number and to be scratched by those with a higher number. Once this framework is established it becomes possible to place all other minerals on the scale by means of the same scratching procedure.

The common touch *left*
Also shown here are four possible scratching implements that can serve as a guide to comparative hardnesses on Mohs' scale.
a Fingernail (Mohs' 2–2.5)
b Penny (4)
c Knife blade (5–6)
d Knife sharpener (8–9)

Knoop hardness scale
right Here we demonstrate another scale of hardness, and compare it with Mohs' scale. Whereas Mohs' scale shows only relative hardness, the Knoop scale gives absolute measurements. Readings on this scale are made by measuring the size of indentation made by a diamond-shaped device dropped on the material. This method makes it possible to give a very precise hardness value to different substances. As is the case with Mohs' scale, the higher the number the harder the substance, but, as our diagram shows, intervals between minerals are now very different. Minerals with values 1–7 on Mohs' scale are concentrated below 1000 on the Knoop scale, 8 and 9 fall below 2000, whereas diamond scores an amazing 7000, taking it right off our diagram!

Hardness and structure
left Here we compare two substances whose relative hardness is due solely to their atomic structure. Diamond, the hardest known natural substance, and graphite, one of the softest minerals, are two different forms or "allotropes" of the same element: carbon.
A In diamond, the carbon atoms form a strong, rigid, crystalline structure.
B In graphite, the atoms are arranged in planes or "sheets." This looser structure gives graphite its slippery character, since the sheets slide easily one over another.

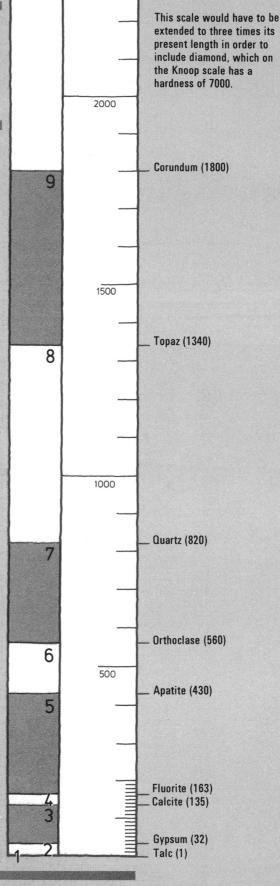

This scale would have to be extended to three times its present length in order to include diamond, which on the Knoop scale has a hardness of 7000.

2000

Corundum (1800)

1500

Topaz (1340)

1000

Quartz (820)

Orthoclase (560)

500

Apatite (430)

Fluorite (163)
Calcite (135)

Gypsum (32)
Talc (1)

Don't be misled into thinking that diamond (10 on Mohs' scale) is twice as hard as apatite (5 on the same scale). If the same two minerals are measured on the Knoop scale, diamond has a hardness of 7000, compared with only 430 for apatite.

THE PLANETS

On these pages we compare the composition, mass and density of the planets, using information based on theories of planetary formation and on the investigation of other factors such as magnetic and gravitational fields. As space exploration proceeds, our theories about other planets will be tested scientifically.

Planet	Mass	Density
Mercury	0.331×10^{27}g	5.4g/cm^3
Venus	4.870×10^{27}g	5.2g/cm^3
Earth	5.976×10^{27}g	5.518g/cm^3
Mars	0.642×10^{27}g	3.95g/cm^3
Jupiter	1899.350×10^{27}g	1.34g/cm^3
Saturn	568.598×10^{27}g	0.7g/cm^3
Uranus	86.891×10^{27}g	1.2g/cm^3
Neptune	102.966×10^{27}g	1.7g/cm^3
Pluto	(?) 1.016×10^{27}g	(?)

Mass and density of the planets *above* This table gives the mass and the density of the planets, listing them in distance order from the Sun. Jupiter is over twice as massive as all the others put together. Earth is the densest planet.

Inside the planets *left*
This diagram allows us to compare the composition of the planets. Each planet is represented as a quarter slice, drawn to scale. A look at their major constituents (see key *far left*) shows two groups of four: the small, dense inner planets with a high iron content; and the large outer planets of low density and high hydrogen content. Pluto's constituents are unknown.

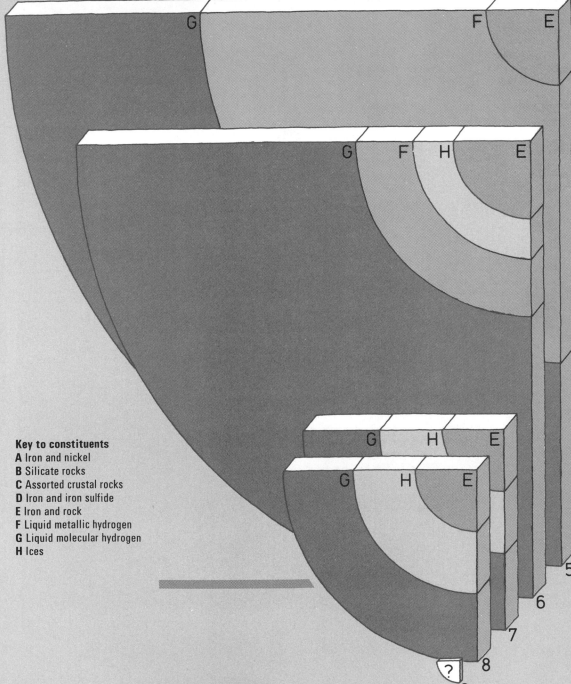

Key to constituents
A Iron and nickel
B Silicate rocks
C Assorted crustal rocks
D Iron and iron sulfide
E Iron and rock
F Liquid metallic hydrogen
G Liquid molecular hydrogen
H Ices

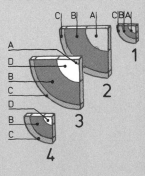

Key to planets
1 Mercury
2 Venus
3 Earth
4 Mars
5 Jupiter
6 Saturn
7 Uranus
8 Neptune
9 Pluto

Among the planets of the solar system, Earth ranks only fifth in terms of mass and of volume, but first in terms of density.

Density	
A Earth	1.00
B Mercury	0.98
C Venus	0.94
D Mars	0.72
E Neptune	0.31
F Jupiter	0.24
G Uranus	0.22
H Saturn	0.13
I Pluto	(?)

Comparative densities
The table *left* and the shaded disks *right* rank the planets in order of density. For easy comparison, the density of Earth—the densest of all the planets—is here taken as "1."

Mass	
F Jupiter	317.83
H Saturn	95.15
E Neptune	17.23
G Uranus	14.54
A Earth	1
C Venus	0.81
I Pluto	0.17(?)
D Mars	0.11
B Mercury	0.06

Comparative masses
The table *left* ranks the planets in order of mass, with the mass of Earth taken as "1." The rows of blocks *right* allow us to compare the planets' masses alongside their densities, and serve to illustrate the difference in their ranking.

Region	Mass	Density
a Crust	2.014×10^{25} g	2.8 g/cm^3
b Mantle	415.347×10^{25} g	4.75 g/cm^3
c Outer core	166.749×10^{25} g	11.1 g/cm^3
d Inner core	13.245×10^{25} g	12.75 g/cm^3

A
B
C
D
E
F
G
H
I

© DIAGRAM

Inside the Earth
The diagram *right* shows the four regions of Earth's interior; the table *left* lists their masses and densities. Knowledge of the crust is based on direct observation; information about the interior is from studies of seismic waves.

a
b
c
d

Earth's crust *left, below*
About 90 elements make up Earth's crust, but over 98% of it is accounted for by just eight of these. The slices in our diagram show the proportions of these first eight elements. The ninth slice, which represents all the other crustal elements,

has been extended to show the proportions of a further eight elements. For the proportions of gold and platinum (five parts per billion) to appear on our diagram as 10mm slivers it would be necessary to increase the circle's radius to 1.978mi (3.183km).

1	Oxygen (O)	46.60%
2	Silicon (Si)	27.72%
3	Aluminum (Al)	8.13%
4	Iron (Fe)	5.00%
5	Calcium (Ca)	3.63%
6	Sodium (Na)	2.83%
7	Potassium (K)	2.59%
8	Magnesium (Mg)	2.09%
9	Titanium (Ti)	0.44%
10	Hydrogen (H)	0.14%
11	Phosphorus (P)	0.12%
12	Manganese (Mn)	0.10%
13	Fluorine (F)	0.08%
14	Sulfur (S)	0.05%
15	Chlorine (Cl)	0.04%
16	Carbon (C)	0.03%
17	Others	0.41%

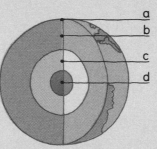

9 10 11 12 13 14 15 16 17

GRAVITY AND WEIGHT

All objects are attracted to each other by a force known as "gravity." The degree of gravitational attraction increases with the mass of an object, and thus it follows that objects within Earth's gravitational pull are attracted to Earth more than to each other. Here we investigate gravity on Earth, in the solar system and in the universe.

Travel to reduce *left*
A person's weight on Earth is the measure of the gravitational pull of Earth's mass on the mass of his body. The farther he is from the center of Earth's mass, the less its pull and the less he weighs. A person on the Equator, where Earth bulges, weighs 0.5% less than at the Poles, where the globe is flattened. Thus a person who weighs 162lb at the North Pole (**a**) loses 13oz by traveling to the Equator (**b**). There is also a 0.5% weight loss for every 6.6mi (10.7km) ascended above sea level (**c**).

How high can you jump?
The table *below* and the diagram *right* show the effect of gravity on the height you can jump. The greater the gravitational force, the more difficult it is to make a high jump. Here we take a jump on Earth of 3ft and show how high a similar jump would be if made elsewhere in the solar system. In the table, gravity is given as the acceleration of a falling object (expressed in feet per second per second), and then, in brackets, relative to that of Earth, which is taken as "1."

	Venue	Gravity		Height jumped
1	Sun	898.88ft/s^2	(27.90)	1¼in
2	Jupiter	75.39ft/s^2	(2.34)	1ft 3½in
3	Neptune	38.02ft/s^2	(1.18)	2ft 6½in
4	Uranus	37.69ft/s^2	(1.17)	2ft 6¾in
5	Saturn	37.04ft/s^2	(1.15)	2ft 7¼in
6	Earth	32.22ft/s^2	(1.00)	3ft 0in
7	Venus	28.35ft/s^2	(0.88)	3ft 4¾in
8	Mars	12.24ft/s^2	(0.38)	7ft 10¾in
9	Mercury	11.91ft/s^2	(0.37)	8ft 1¼in
10	Moon	5.31ft/s^2	(0.16)	18ft 9in

The world record for an overhead lift is 564¼lb. A similar effort on the Moon, where gravity is less, would have resulted in a lift of 3526½lb— equal to lifting 2½ Mini cars.

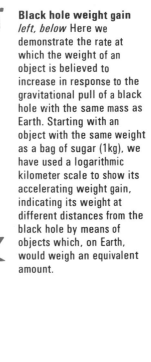

Black hole weight gain
left, below Here we demonstrate the rate at which the weight of an object is believed to increase in response to the gravitational pull of a black hole with the same mass as Earth. Starting with an object with the same weight as a bag of sugar (1kg), we have used a logarithmic kilometer scale to show its accelerating weight gain, indicating its weight at different distances from the black hole by means of objects which, on Earth, would weigh an equivalent amount.

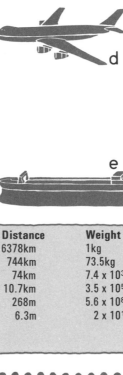

Bending the rules *above*
A Here we represent space as a flat surface with light and time running in straight lines across it.
B The great mass of a star exerts such a strong gravitational force that it pulls these lines toward itself, bending the lines and producing the effect of a dent in the curved surface. Since the lines representing light and time are now longer, it follows that the passage of light and time will be slowed down.

C In this representation of a ''black hole'' our original flat surface is distorted to the extreme and is pulled down to a point in the center. A black hole is thought to result from the collapse of a heavy star under the force of its own gravity. The same amount of matter occupies a smaller and smaller volume, and the force of gravity that it exerts becomes so intense that it is impossible even for light to escape from it, so making it appear black.

Distance	Weight	Equivalent
6378km	1kg	**a** Bag of sugar
744km	73.5kg	**b** Average US man
74km	7.4×10^3kg	**c** Tractor
10.7km	3.5×10^5kg	**d** Jumbo Jet
268m	5.6×10^8kg	**e** Largest oil tanker
6.3m	2×10^{12}kg	**f** World population

©DIAGRAM

This black dot shows the calculated size of a black hole with the same mass as Earth.

COMPARATIVE WEIGHTS

Even when objects are small enough to hold in the hands it is by no means easy to estimate comparative weights with any degree of accuracy. With very heavy objects the problem is even greater. One useful guide is, as here, to measure unfamiliar weights in terms of other, generally more familiar ones.

Tons of bricks *below*
Walls built from a ton of standard bricks show considerable variations in size depending on the type of ton—US (short), UK/imperial (long), or metric. Here we compare the lengths of three walls built from a "ton" of bricks. In each case the wall is 4ft high and built from bricks measuring, with mortar, 3 x 4½ x 9in.
a Wall from a UK/imperial ton of bricks 39ft.
b Wall from a metric tonne of bricks 38ft 3in.
c Wall from a US ton of bricks 34ft 6in.

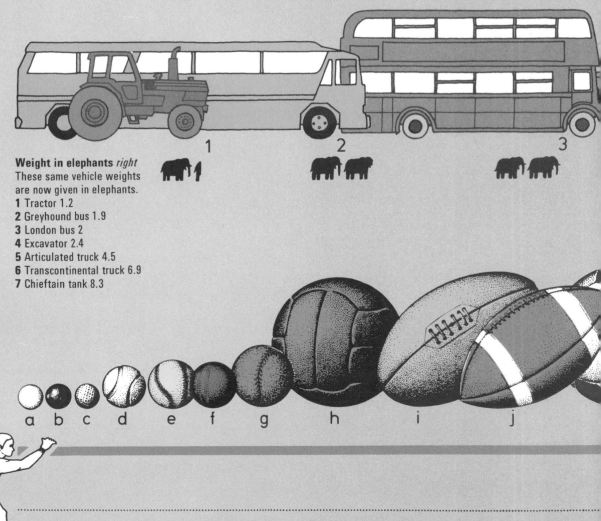

Vehicular heavyweights
right These heavy land vehicles have all been drawn to a common scale. Weights are given here in short tons (each 2000 lb).
1 Tractor 8.2 tons.
2 American Greyhound bus (model MC8) 13.4 tons.
3 London double-decker bus (model RML) 13.8 tons.
4 Excavator 16.5 tons.
5 Articulated truck (with load) 31.3 tons.
6 Transcontinental truck (with load) 48.5 tons.
7 Chieftain tank, the heaviest tank in current use, 58.2 tons.

Weight in elephants *right*
These same vehicle weights are now given in elephants.
1 Tractor 1.2
2 Greyhound bus 1.9
3 London bus 2
4 Excavator 2.4
5 Articulated truck 4.5
6 Transcontinental truck 6.9
7 Chieftain tank 8.3

Fair weights for balls
Shown to scale *right* and listed in the table *far right* in order of weight, from the lightest to the heaviest, are officially approved balls for selected sporting activities.

a b c d e f g h i j

One stone at Stonehenge is equal in weight to 37,200 standard bricks. These bricks (each $3 \times 4\frac{1}{2} \times 9$in) would be sufficient to build a one brick thick chimney with a diameter of 12ft and a height of 186ft.

Weighty stones *right*
Limestone blocks used for the Great Pyramid of Cheops (**1**) weigh an average 2.8 short tons, equal to four Minis (each 1406 lb). Sarsen blocks at England's Stonehenge (**2**) weigh over 50 short tons each, equal to over 71 Minis.

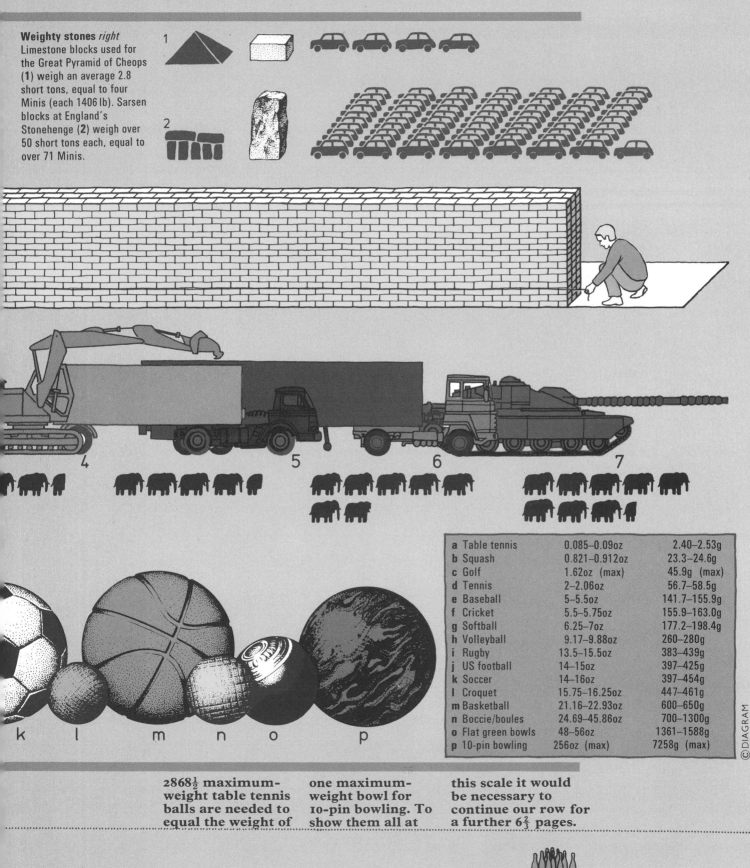

a	Table tennis	0.085–0.09oz	2.40–2.53g
b	Squash	0.821–0.912oz	23.3–24.6g
c	Golf	1.62oz (max)	45.9g (max)
d	Tennis	2–2.06oz	56.7–58.5g
e	Baseball	5–5.5oz	141.7–155.9g
f	Cricket	5.5–5.75oz	155.9–163.0g
g	Softball	6.25–7oz	177.2–198.4g
h	Volleyball	9.17–9.88oz	260–280g
i	Rugby	13.5–15.5oz	383–439g
j	US football	14–15oz	397–425g
k	Soccer	14–16oz	397–454g
l	Croquet	15.75–16.25oz	447–461g
m	Basketball	21.16–22.93oz	600–650g
n	Boccie/boules	24.69–45.86oz	700–1300g
o	Flat green bowls	48–56oz	1361–1588g
p	10-pin bowling	256oz (max)	7258g (max)

©DIAGRAM

$2868\frac{1}{2}$ maximum-weight table tennis balls are needed to equal the weight of one maximum-weight bowl for 10-pin bowling. To show them all at this scale it would be necessary to continue our row for a further $6\frac{2}{3}$ pages.

AIRCRAFT

With its pilot, the Wright Brothers' aircraft *Flyer 1* weighed only 750lb, less than five times the weight of an average US man (162lb). The wide-bodied Boeing 747 "Jumbo Jet" has a maximum take-off weight of 775,000lb, equal to the weight of 4784 average men or 55 times the weight of an average African elephant (14,000lb).

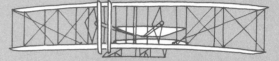

Much heavier than air
Listed from the lightest to the heaviest in the table *right*, and illustrated and plotted by weight on the logarithmic scale *below* is a selection of aircraft, both civil and military, old and new. The weight given for *Flyer I* includes the weight of the pilot. All other weights are maximum take-off weights.

			Civil aircraft
750 lb	340kg	A	Wright brothers' *Flyer I*, first flown in 1903
1600 lb	726kg	B	Beagle B121 Pup-150, 2/3 seat cabin monoplane
25,000 lb	11,340kg	C	HS 125 Series 600, light executive transport
333,600 lb	151,321kg	D	Boeing 707-320B, long-range airliner
408,000 lb	185,069kg	E	BAC/Aérospatiale Concorde, supersonic transport
775,000 lb	351,540kg	F	Boeing 747 "Jumbo Jet," wide-bodied long-range airliner
			Military aircraft
1453 lb	659kg	G	Sopwith F1 Camel, WW1 Allied fighter
7700 lb	3493kg	H	Messerschmitt Bf 109G-X, WW2 German fighter
8745 lb	3967kg	I	Gotha GVb, WW1 German bomber
56,000 lb	25,402kg	J	McDonnell Douglas F-15 Eagle, modern fighter
68,000 lb	30,845kg	K	Avro Lancaster B1, WW2 British bomber
488,000 lb	221,357kg	L	Boeing B-52H, heaviest modern bomber
672,000 lb	304,819kg	M	British Hovercraft Corporation SRA4 Mark 3
6,526,000 lb	2,960,194kg	N	Saturn V rocket

Bicycles to airplanes
above Before turning to making airplanes, Orville and Wilbur Wright were bicycle makers. *Flyer I*, in which Orville made the first controlled and sustained flight ever, weighed 605 lb when empty—20 times the weight of a modern touring bicycle.

Civil aircraft *right*
Selected civil aircraft, listed in the table *above*, are here plotted against a logarithmic weight scale. Except for the pioneering *Flyer I*, all of the civil planes shown are in current use.

Military aircraft *right*
Included here are representative fighters and bombers from the two world wars and from the present day. The modern fighter is similar in weight to a WW2 bomber; a WW2 fighter is similar in weight to a WW1 bomber.

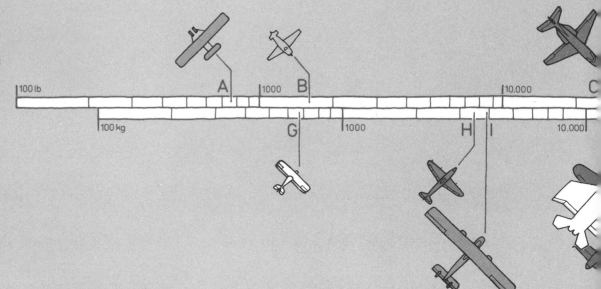

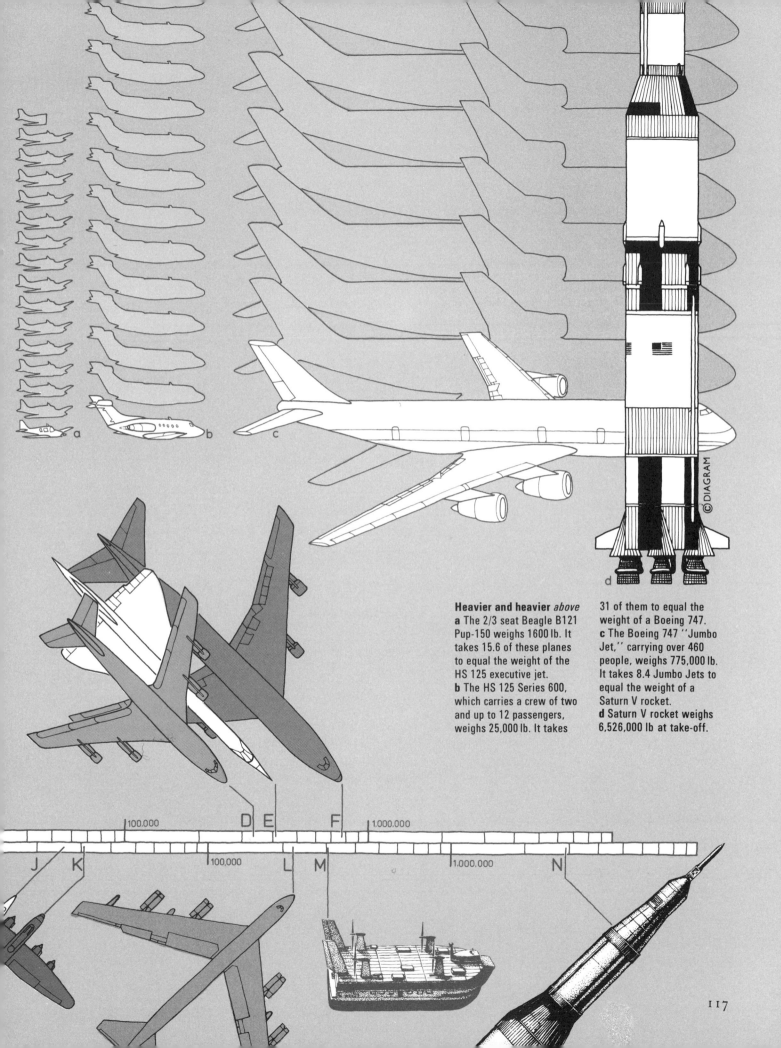

Heavier and heavier *above*
a The 2/3 seat Beagle B121 Pup-150 weighs 1600 lb. It takes 15.6 of these planes to equal the weight of the HS 125 executive jet.
b The HS 125 Series 600, which carries a crew of two and up to 12 passengers, weighs 25,000 lb. It takes 31 of them to equal the weight of a Boeing 747.
c The Boeing 747 "Jumbo Jet," carrying over 460 people, weighs 775,000 lb. It takes 8.4 Jumbo Jets to equal the weight of a Saturn V rocket.
d Saturn V rocket weighs 6,526,000 lb at take-off.

100,000   D E   F   1,000,000

100,000   L M   1,000,000   N

J   K

ANIMALS

The weights of living creatures range from the infinitesimal weights of microscopic organisms to the tons of an elephant or whale. The blue whale weighs over 70 million times more than the smallest bird, Helena's hummingbird, and 24,560 billion times more than the smallest insect, the parasitic wasp (0.0000002oz).

Farmyard equivalents *left, below* Here we show a farmer's wife—a US woman of average weight (135 lb)—and show how many chickens, cats and rats make up the same weight.
a Chickens 19.3
b Cats 9.6
c Rats 134.6

Creatures great and small Plotted on the logarithmic scales *right* and listed in the table *far right* are likely average weights for animal species, from the smallest bird to the heaviest living creature, the blue whale. Also included for comparison is a *Brachiosaurus*, the heaviest known dinosaur. Average weights for animal species are always subject to some uncertainty, especially for wild animals. Man may not have weighed, or even seen, the heaviest or lightest of a species, and big seasonal variations in weight are a further complicating factor.

Scale markings (top to bottom): 10,000g · 10 lb · 1,000g · 1 lb · 10 oz · 100g · 1 oz · 0.5 oz · 10g · 0.1 oz · 0.05 oz · 1g

37,029 Helena's hummingbirds (shown *left* actual size) equal the weight of an average man; 30,858 equal the weight of an average woman. To represent these numbers of Helena's hummingbirds by symbols similar to those *above* we would need 1252 pages for a man's weight, and 1043 for a woman's.

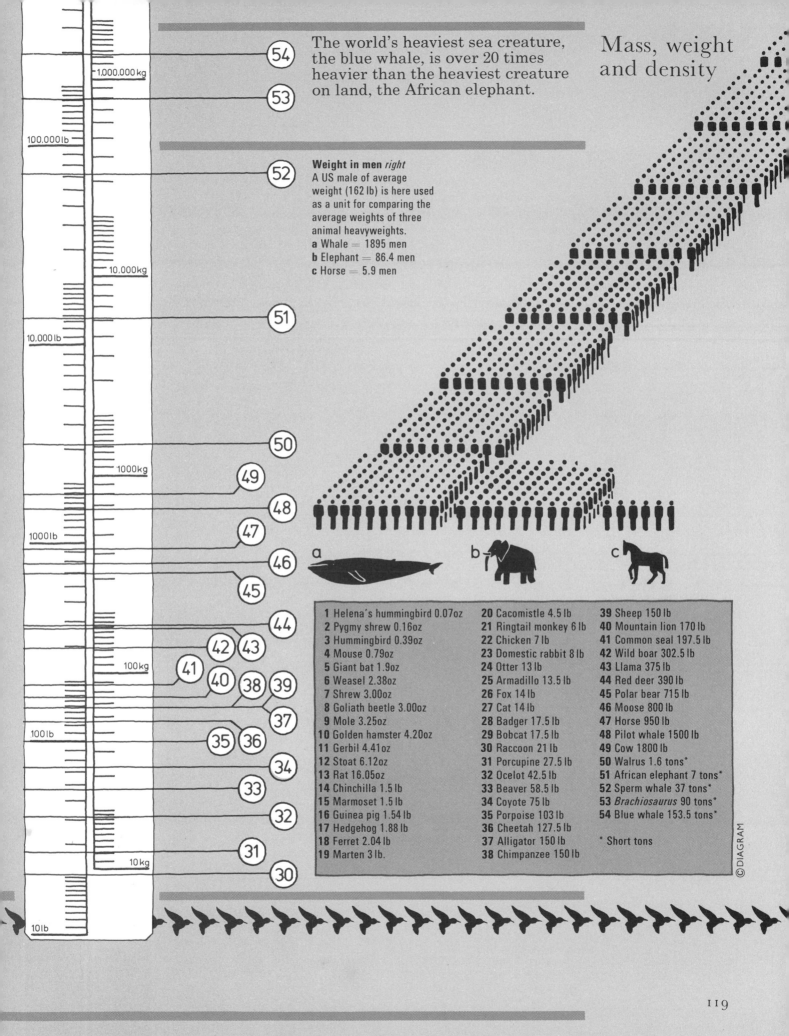

The world's heaviest sea creature, the blue whale, is over 20 times heavier than the heaviest creature on land, the African elephant.

Weight in men *right*
A US male of average weight (162 lb) is here used as a unit for comparing the average weights of three animal heavyweights.
a Whale = 1895 men
b Elephant = 86.4 men
c Horse = 5.9 men

1 Helena's hummingbird 0.07oz
2 Pygmy shrew 0.16oz
3 Hummingbird 0.39oz
4 Mouse 0.79oz
5 Giant bat 1.9oz
6 Weasel 2.38oz
7 Shrew 3.00oz
8 Goliath beetle 3.00oz
9 Mole 3.25oz
10 Golden hamster 4.20oz
11 Gerbil 4.41oz
12 Stoat 6.12oz
13 Rat 16.05oz
14 Chinchilla 1.5 lb
15 Marmoset 1.5 lb
16 Guinea pig 1.54 lb
17 Hedgehog 1.88 lb
18 Ferret 2.04 lb
19 Marten 3 lb.

20 Cacomistle 4.5 lb
21 Ringtail monkey 6 lb
22 Chicken 7 lb
23 Domestic rabbit 8 lb
24 Otter 13 lb
25 Armadillo 13.5 lb
26 Fox 14 lb
27 Cat 14 lb
28 Badger 17.5 lb
29 Bobcat 17.5 lb
30 Raccoon 21 lb
31 Porcupine 27.5 lb
32 Ocelot 42.5 lb
33 Beaver 58.5 lb
34 Coyote 75 lb
35 Porpoise 103 lb
36 Cheetah 127.5 lb
37 Alligator 150 lb
38 Chimpanzee 150 lb

39 Sheep 150 lb
40 Mountain lion 170 lb
41 Common seal 197.5 lb
42 Wild boar 302.5 lb
43 Llama 375 lb
44 Red deer 390 lb
45 Polar bear 715 lb
46 Moose 800 lb
47 Horse 950 lb
48 Pilot whale 1500 lb
49 Cow 1800 lb
50 Walrus 1.6 tons*
51 African elephant 7 tons*
52 Sperm whale 37 tons*
53 *Brachiosaurus* 90 tons*
54 Blue whale 153.5 tons*

* Short tons

©DIAGRAM

HUMANS

People's individual weights vary within the limits set by our species. The heaviest person, for instance, was over nine times heavier than is average for a US adult, and over 12 times as heavy as a flyweight boxer. Here we present some usual and unusual aspects of weight within the human species.

His and her weights *below*
The diagram shows the average weights of US males and females at ages from birth to 65 years. The average weight for a US adult male is 162 lb and for an adult female 135 lb. Men reach their maximum weight, an average of around 173 lb, between the ages of 35 and 54 years. Women reach their greatest weight, an average 152 lb, at around 55–60 years.

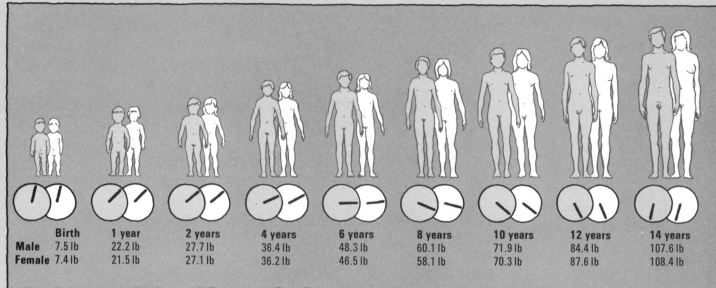

	Birth	1 year	2 years	4 years	6 years	8 years	10 years	12 years	14 years
Male	7.5 lb	22.2 lb	27.7 lb	36.4 lb	48.3 lb	60.1 lb	71.9 lb	84.4 lb	107.6 lb
Female	7.4 lb	21.5 lb	27.1 lb	36.2 lb	46.5 lb	58.1 lb	70.3 lb	87.6 lb	108.4 lb

Boxing and wrestling *right*
The narrow weight divisions of boxing and freestyle wrestling attempt to ensure that the contest is as fair as possible. The weights given here are the official weight limits for the AIBA (Amateur International Boxing Association), the WBC (World Boxing Council), and, *far right*, for Olympic wrestling. The weights for the AIBA and for Olympic wrestling are officially in kilograms. WBC weights are officially in pounds.

Class	AIBA	WBC	Wrestling
Light flyweight	48kg (105.8 lb)	108 lb (49.0kg)	48kg (105.8 lb)
Flyweight	51kg (112.4 lb)	112 lb (50.8kg)	52kg (114.6 lb)
Bantamweight	54kg (119.0 lb)	118 lb (53.5kg)	57kg (125.7 lb)
Super bantamweight	– –	122 lb (55.3kg)	– –
Featherweight	57kg (125.7 lb)	126 lb (57.2kg)	62kg (136.7 lb)
Junior lightweight	– –	130 lb (59.0kg)	– –
Lightweight	60kg (132.3 lb)	135 lb (61.2kg)	68kg (149.9 lb)
Light welterweight	63.5kg (140.0 lb)	140 lb (63.5kg)	– –
Welterweight	67kg (147.7 lb)	147 lb (66.7kg)	74 kg (163.1 lb)
Light middleweight	71kg (156.5 lb)	154 lb (69.9kg)	– –
Middleweight	75kg (165.3 lb)	160 lb (72.6kg)	82kg (180.8 lb)
Light heavyweight	81kg (178.6 lb)	175 lb (79.4kg)	90kg (198.4 lb)
Heavyweight	81+ kg (178.6+ lb)	175+ lb (79.4+ kg)	100kg (220.5 lb)
Heavyweight plus	– –	– –	100+ kg (220.5+ lb)

Your weight in gold?
right It is said that kings used to receive their tributes in an amount equal to their own weights. The average US man weighs 162 lb. Shown here are the volumes that would be taken up by the same weight of various solids and liquids.

1	Gold	0.14ft³
2	Copper	0.29ft³
3	Diamond	0.74ft³
4	Glass	1.00ft³
5	Sugar	1.62ft³
6	Coal	1.92ft³
7	Nylon	2.33ft³
8	Milk	2.52ft³
9	Sea water	2.53ft³
10	Gasoline	3.86ft³

Perhaps humans have an excuse for forgetting things— our brains weigh about 3lb, only one fifth as much as an elephant's brain.

A six-year-old child weighs roughly the same as the air in a furnished bedroom measuring 9 × 9 × 8ft.

An average man's weight is 21.6 times more than the average weight for a newborn boy baby. An average woman weighs 18.2 times the weight of an average newborn girl.

Heavyweight humans *right*
The weights of the heaviest humans are here compared with average weights for US men and women, and with typical weights of selected animals.

a Jon Brower Minnoch (born 1941), man with the highest estimated weight.
b Robert Earl Hughes (1926–58), man with the highest undisputed weight.

c Ida Maitland (1898–1932), woman for whom the highest weight has been claimed.
d Mrs Percy P. Washington (1926–72), woman with the greatest weight registered on scales (800 lb); total weight estimated at 880 lb.

16 years	18 years	25 years	45 years	65 years
129.7 lb	143.0 lb	153.0 lb	173.0 lb	164.0 lb
117.0 lb	120.0 lb	124.0 lb	139.0 lb	134.0 lb

© DIAGRAM

An average girl reaches 50% of her adult weight at age nine. An average boy does not reach 50% of his adult weight until he is eleven.

1 2 3 4 5 6 7 8 9 10

kg / lb

● **a** Minnoch 1400 lb
650

Sea cow 1300 lb ●
1250

550

Saltwater crocodile 1100 lb ●
● **b** Hughes 1069 lb
1000

450

Horse 950 lb ●
● **c** Maitland 911 lb
● **d** Washington 880 lb

350

Polar bear 715 lb ●
750

250

Gorilla 450 lb ●
500

150

250

Chimpanzee 150 lb ●
● Average US man 162 lb
● Average US woman 135 lb

50

WEIGHTS LIFTED

Individual strength depends on many personal factors such as size, weight and muscular development. Most normally healthy people should be able to lift about their own weight. Here some of the heavier weights lifted by men and women are translated into more easily understandable terms for us weaker mortals!

Digital strength *right*
Warren L. Travis is reported to have lifted 667 lb with one finger. This is equivalent to lifting four men of average weight (162 lb).

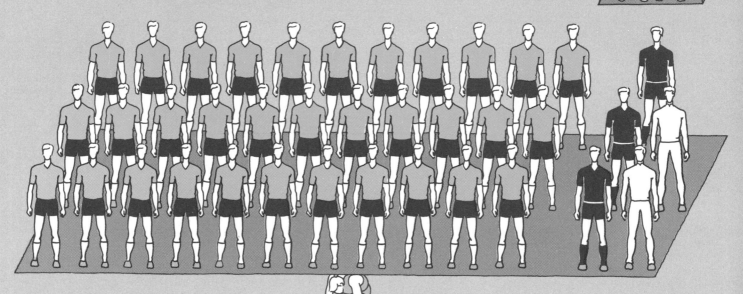

Back to basics *above*
The greatest weight ever lifted by a man was 6270 lb, in a backlift in 1957 by Olympic weightlifting champion Paul Anderson. The weight comprised a table on which there was a safe filled with lead and other pieces of heavy metal.

This is equivalent to lifting:
3 soccer teams;
1 referee;
2 linesmen; and
2 substitutes—
making a total of 38 men of average weight!

Hip lady ! *left*
The heaviest weight ever lifted by a woman was 3564 lb, in a hip and harness lift by Josephine Blatt in 1895. This is equivalent to lifting 26 chorus girls the weight of an average US woman (135 lb).

Light as air? *left*

A column of air with a 1in² base weighs 14.7 lb at sea level. This means that as we walk on the beach we are carrying over 1000 lb of air on our heads and shoulders.

Family feats *right, below*

The world record for an overhead lift is a jerk of 564¼ lb made by Vasili Alexeev in 1977. This is equivalent to lifting (**a**) a family comprising father (162 lb), mother (135 lb), boys aged 14 (108 lb) and 10 (72 lb), and girls aged 6 (48 lb) and 4 (36 lb). The heaviest overhead lift made by a woman is a continental jerk of 286 lb made by Katie Sandwina in c.1911. Her record is comparable with lifting (**b**) a husband (162 lb), a boy of 10 (72 lb) and a girl of 6 (48 lb).

a

b

An ant can lift 50 times its own body weight. If an average man could do the same, he would be able to lift 8100lb—29% more than the existing record.

©DIAGRAM

CHAPTER 6

This illustration from
Diderot's *Encyclopédie* of
1751 shows one man
working an enormous wine
press. Mechanical ingenuity
can be used to compensate
for a limited supply of power.

A steam-generating heavy
water reactor in Dorset,
England. Nuclear power may
solve the energy crisis, but
strict safety regulations
are vital (Photo: UK Atomic
Energy Authority).

ENERGY

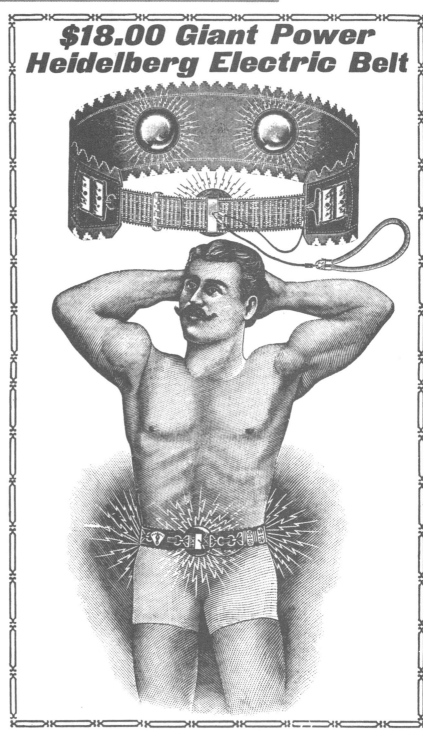

Illustration from the Sears Roebuck catalogue of 1902 showing an electric belt claimed as a cure for ''disorders of the nerves, stomach, liver and kidneys,'' and for ''weakness, diseased or debilitated condition of the sexual organs from any cause whatever. . .''

MEASURING ENERGY

b

On these two pages we look at different ways of measuring energy, and on subsequent pages we consider some of the forms that energy takes. Energy is defined as the equivalent of or the capacity to do work. Power is a measure of the amount of energy spent over a certain period of time. (Temperature comparisons are in chapter 7.)

Unit equivalents
$1J = 10^7 erg$
$1erg = 10^{-7}J$
$1cal = 4.1855J$
$1000cal = 1$ kilocalorie, or 1 Calorie
$1Btu = 1054.5J$
$1W = 0.001341hp$
$1hp = 745.7W$
1 hp (metric) $= 735.5W$

Energy and power units
The table *right* defines the major units of energy and power in use today. As most countries measure their electricity in watts, we take the joule and the watt as our basic units. Equivalent values for units are listed in the small table *left*.

The joule (J) is the amount of energy needed to move a mass of one kilogram through one meter with an acceleration of one meter per second per second.
The erg is the amount of energy needed to move one gram through one centimeter with an acceleration of one centimeter per second per second.
The calorie (cal, or more specifically, cal_{15}) is the amount of energy needed to raise the temperature of one gram of water by one degree Celsius (or Centigrade) from 14.5°C to 15.5°C (58.1°F to 59.9°F).
The British thermal unit (Btu) is the energy needed to raise the temperature of one pound of water from 60°F to 61°F (15.5°C to 16.1°C).
The watt (W) is the power provided when one joule is used for one second. 1000 watts are known as a **kilowatt** (kW).
The kilowatt hour (kWh) is the energy expended when one kilowatt is available for one hour.
The British horsepower (hp) is the power needed to raise 550 lb one foot in one second.
The metric horsepower is the power needed to raise 75kg one meter in one second.

	1			2		3		4		5
A	10^{41}	10^{40}J			10^{35}			10^{30}		10^{25}
B	10^{48}	erg		10^{45}		10^{40}		10^{35}		

A	10^7J	10^6	10^5	10^4	10^3	10^2	10	1	10^{-1}
B	erg				10^{10}	10^9	10^8	10^7	10^6
C	10^7cal	10^6	10^5	10^4	10^3	10^2	10	1	D 10^{18}e

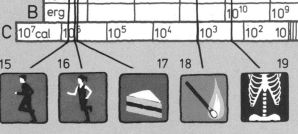

| | 15 | 16 | | 17 | 18 | 19 | 20 |

Universal energy scales *above, right* Here joules and ergs appear on equivalent logarithmic scales that extend, left to right on two bars, from 10^{41} to 10^{-19}J (scale **A**), and from 10^{48} to 10^{-12}erg (scale **B**). These scales encompass the whole range of energy known to man. If we were to show the same energy range on a conventional scale with one erg equal to one millimeter, our scale bar would be some 10^{42}km long !

Included below our joule and erg scales are equivalent scales for calories (scale **C**) and electronvolts (scale **D**). These scales cover the energy ranges over which these units are normally used. The electronvolt (eV) is a very small unit used in atomic and subatomic physics. One electronvolt is equal to 1.602×10^{-19}J. The energy contents of selected phenomena are plotted on the scales and listed *right*.

Energies compared *above*
1 Supernova 10^{41}J
2 Annual energy output of the Sun 10^{34}J
3 Earth traveling in orbit around the Sun 2×10^{33}J
4 Earth spinning on its axis 3×10^{29}J
5 Earth's annual share of solar radiation 5.6×10^{24}J

6 Eruption (1883) of the volcano Krakatoa 6×10^{18}J
7 Very severe earthquake (Richter 8) 10^{18}J
8 100Mt thermonuclear bomb 4.2×10^{17}J
9 Hurricane 4×10^{15}J
10 Atomic bomb dropped on Hiroshima 8.4×10^{13}J

11 Energy generated by Saturn V rocket 1.3×10^{11}J
12 Explosion of 1 short ton of TNT 4.2×10^9J
13 A day's heavy manual labor 1.7×10^7J
14 Energy content of 1 lb of good quality bituminous coal 1.5×10^7J

The energy in a slice of apple pie (350 Calories) is equivalent to the energy content of 11.2oz of TNT. A young man's 3000 Calories of food energy eaten daily is equivalent to the energy in 6 lb of TNT.

The energy content of 1 lb of good quality bituminous coal is 3750 times the amount of energy used up by a burning match.

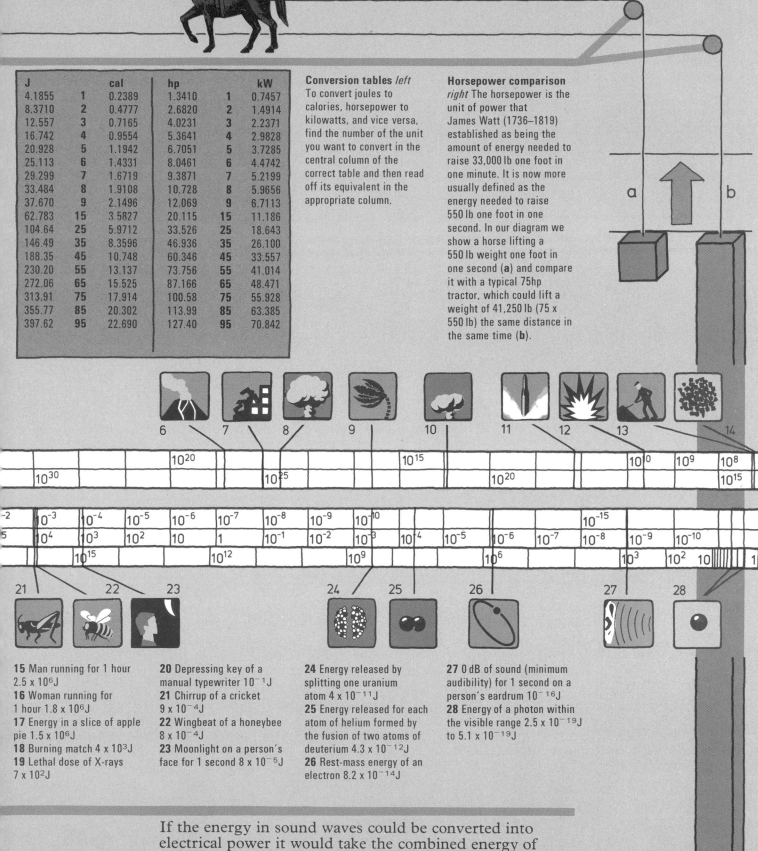

J		cal	hp		kW
4.1855	1	0.2389	1.3410	1	0.7457
8.3710	2	0.4777	2.6820	2	1.4914
12.557	3	0.7165	4.0231	3	2.2371
16.742	4	0.9554	5.3641	4	2.9828
20.928	5	1.1942	6.7051	5	3.7285
25.113	6	1.4331	8.0461	6	4.4742
29.299	7	1.6719	9.3871	7	5.2199
33.484	8	1.9108	10.728	8	5.9656
37.670	9	2.1496	12.069	9	6.7113
62.783	15	3.5827	20.115	15	11.186
104.64	25	5.9712	33.526	25	18.643
146.49	35	8.3596	46.936	35	26.100
188.35	45	10.748	60.346	45	33.557
230.20	55	13.137	73.756	55	41.014
272.06	65	15.525	87.166	65	48.471
313.91	75	17.914	100.58	75	55.928
355.77	85	20.302	113.99	85	63.385
397.62	95	22.690	127.40	95	70.842

Conversion tables *left*
To convert joules to calories, horsepower to kilowatts, and vice versa, find the number of the unit you want to convert in the central column of the correct table and then read off its equivalent in the appropriate column.

Horsepower comparison *right* The horsepower is the unit of power that James Watt (1736–1819) established as being the amount of energy needed to raise 33,000 lb one foot in one minute. It is now more usually defined as the energy needed to raise 550 lb one foot in one second. In our diagram we show a horse lifting a 550 lb weight one foot in one second (**a**) and compare it with a typical 75hp tractor, which could lift a weight of 41,250 lb (75 x 550 lb) the same distance in the same time (**b**).

15 Man running for 1 hour 2.5×10^6J
16 Woman running for 1 hour 1.8×10^6J
17 Energy in a slice of apple pie 1.5×10^6J
18 Burning match 4×10^3J
19 Lethal dose of X-rays 7×10^2J

20 Depressing key of a manual typewriter 10^{-1}J
21 Chirrup of a cricket 9×10^{-4}J
22 Wingbeat of a honeybee 8×10^{-4}J
23 Moonlight on a person's face for 1 second 8×10^{-5}J

24 Energy released by splitting one uranium atom 4×10^{-11}J
25 Energy released for each atom of helium formed by the fusion of two atoms of deuterium 4.3×10^{-12}J
26 Rest-mass energy of an electron 8.2×10^{-14}J

27 0 dB of sound (minimum audibility) for 1 second on a person's eardrum 10^{-16}J
28 Energy of a photon within the visible range 2.5×10^{-19}J to 5.1×10^{-19}J

If the energy in sound waves could be converted into electrical power it would take the combined energy of 100,000,000,000,000,000 (10^{17}) mosquito buzzes to provide enough power to light a reading lamp.

ENERGY SOURCES

The total amount of energy potentially available each year at Earth's surface is more than 20 billion times greater than present energy consumption, but as yet man has learned to make little use of solar radiation, which makes up nearly all of this total. Instead he uses increasingly scarce, non-renewable sources such as oil.

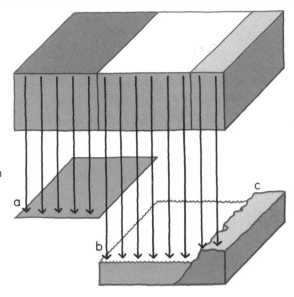

Solar and non-solar *left*
For this diagram we look only at energy that is potentially available at Earth's surface every year (as distinct from "non-renewable" sources such as oil). Of this annual total, the ratio of solar to non-solar energy is 2600:1.

Solar radiation *right*
An estimated 1.56×10^{18} kWh of solar energy (1.95×10^{14} tonnes of coal equivalent) annually reaches the outer limit of Earth's atmosphere. Of this: 40% is absorbed by the atmosphere (**a**); 44% reaches sea areas (**b**); 16% reaches land areas (**c**).

	1968	**1978**
1 Oil	2892 (42.0%)	4614 (46.0%)
2 Coal	2330 (33.8%)	2717 (27.1%)
3 Natural gas	1233 (17.9%)	1861 (18.6%)
4 Water power	413 (6.0%)	606 (6.0%)
5 Nuclear power	20 (0.3%)	228 (2.3%)
Total of above	6888 (100%)	10,026 (100%)

Figures in million metric tonnes of coal equivalent

Consumption comparisons
The table *left* provides a comparison of the five most important sources of world primary energy consumption in 1968 and 1978 (based on information in the BP Statistical Review of the World Oil Industry, 1978). Consumption of energy from each of these sources increased from 1968 to 1978. Although there was no change in their order of importance, the percentage share of the annual total increased noticeably for oil and nuclear power and decreased for coal. 1978 percentages are shown *below*.

Consumption increases
right Levels of primary energy consumption from selected sources in 1978 are here shown as multiples of their 1968 equivalents.
a Total of five major sources (x 1.6)
b Nuclear power (x 11.4)
c Coal (x 1.2)

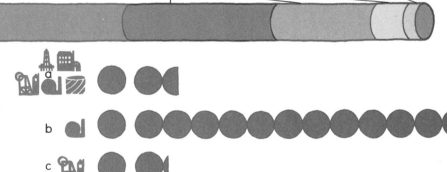

The energy equivalent of 1 day's consumption of electricity in the United States would be sufficient to power an automobile 36,000 times around the world.

The Middle East produces more than 10 times as much energy as it consumes. Western Europe consumes over twice as much as it produces.

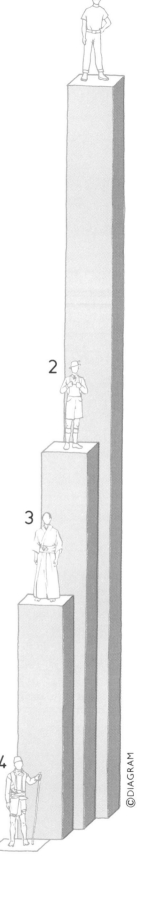

	1 Production	**2 Consumption**
World	8,951,000,000t	8,318,000,000t
a N America	2,309,000,000t	2,716,000,000t
b E Europe/USSR	2,162,000,000t	1,922,000,000t
c Middle East	1,688,000,000t	144,000,000t
d Other Asia	1,020,000,000t	1,351,000,000t
e W Europe	680,000,000t	1,573,000,000t
f Africa	537,000,000t	164,000,000t
g C/S America	429,000,000t	342,000,000t
h Oceania	126,000,000t	106,000,000t

Energy production and consumption The diagram *left* and the table *above* provide a comparison of energy production and consumption in different parts of the world. Figures, collected by the United Nations, are for 1976. Energy totals are expressed in metric tonnes of coal equivalent; 1 tonne of coal can be taken as providing 8000kWh of energy. The Middle East, E Europe and the USSR, Africa, C and S America, and Oceania have energy surpluses; N America (the region producing most energy), W Europe, and the rest of Asia have deficits.

Consumption per person *right* The diagram shows great national differences in per capita consumption. Figures (1976) give kg of coal equivalent per capita:
1 USA 11,554kg
2 W Germany 5922kg
3 Japan 3679kg
4 Nepal 11kg

Fuel efficiencies *below* This diagram, with its key beneath it, shows the relative efficiency of different fuels. Shown are the amounts of water, in liters, that can be boiled by 1 kilogram of each fuel. Natural gas is nearly 10 times as efficient as coal.

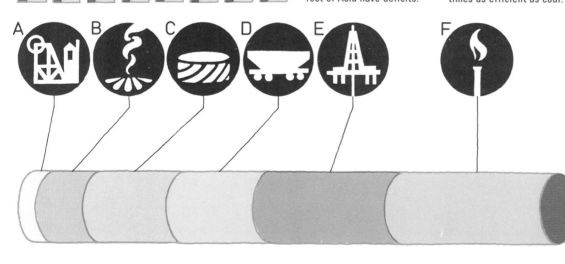

A Coal 2 liters
B Wood 6 liters
C Coal gas 10 liters
D Anthracite 10 liters
E Gasoline 16 liters
F Natural gas 19 liters

To exert power equivalent to the amount of electricity used by US manufacturing industry in a single week, one worker doing average manual labor would have to work for 145,193,740 years (working fifty 40-hour weeks per year).

VOLCANOES, BOMBS AND EARTHQUAKES

The greatest releases of energy on Earth occur during volcanic eruptions, earthquakes and nuclear explosions. The eruption of Tambora in 1815 is believed to have been over 100 times more powerful than the largest nuclear test. The latter, at Novaya Zemlya, was comparable to an earthquake of Richter magnitude 7 to 8.

Scale B (right)
- 100,000 Mt
- 10,000 Mt
- 1000 Mt
- 100 Mt
- 10 Mt
- 1 Mt
- 100 kt
- 10 kt
- 1 kt
- 100 t
- 10 t
- 1 t

Scales of destruction
right, far right Four scales are used here.
Scale A on both diagrams gives energy in joules.
Scale B shows short tons, kilotons (kt = 1000t), and megatons (Mt = 1,000,000t) of TNT equivalent, where one ton equals 4.2×10^9J.
Scale C is Richter's scale for measuring earthquakes.
Scale D is Mercalli's scale for measuring earthquakes.

Scale A

| 10^{21} | 10^{18} | 10^{15} | 10^{12} | 10^{9} | 10^{6} | 10^{3} | 1 |

Volcanoes and bombs
above, right Shown here are estimates of energy from volcanoes and nuclear explosions. For Tambora (**a**), the greatest volcanic eruption known, we give the total release of energy. For Santorini (**b**) and for Krakatoa (**c**) we use figures for the greatest single explosions. The H-bomb tested at Novaya Zemlya (**d**) was the biggest to date; Eniwetok (**e**) was the first H-bomb test. The bomb tested at Alamogordo (**f**) and the bombs dropped on Hiroshima (**g**) and Nagasaki (**h**) were identical A-bombs.

Volcanic eruptions	Joules	TNT equivalent
a Tambora, Indonesia (April 1815)	8×10^{19}J	20,000Mt
b Santorini, Greece (c. 1470BC)	3×10^{19}J	7500Mt
c Krakatoa, Indonesia (August 1883)	6×10^{18}J	1500Mt
Nuclear explosions		
d Novaya Zemlya, USSR (October 1961)	3×10^{17}J	60Mt
e Eniwetok, US Pacific (November 1952)	6×10^{16}J	15Mt
f Alamogordo, USA (July 1945)	8×10^{13}J	20kt
g Hiroshima, Japan (August 1945)	8×10^{13}J	20kt
h Nagasaki, Japan (August 1945)	8×10^{13}J	20kt

A- and H-bombs compared
left Nuclear bombs work in one of two ways. An atomic bomb functions by fission, using the energy released by the splitting of uranium and plutonium atoms. A hydrogen—or thermonuclear—bomb uses the energy released by the fusion of two hydrogen atoms into one helium atom. Our diagram compares the 80,000MW of power produced by one gram of fissioning uranium (**1**), with the 240,000MW of power that one gram of deuterium (an isotope of hydrogen) yields through fusion (**2**).

The H-bomb tested at Novaya Zemlya in October 1961 was 3000 times as powerful as the A-bomb dropped on Hiroshima in August 1945.

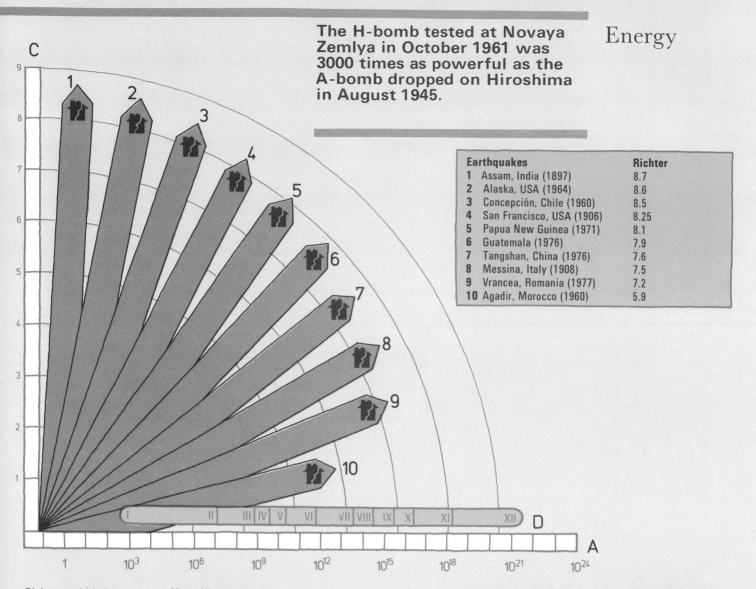

Earthquakes	Richter
1 Assam, India (1897)	8.7
2 Alaska, USA (1964)	8.6
3 Concepción, Chile (1960)	8.5
4 San Francisco, USA (1906)	8.25
5 Papua New Guinea (1971)	8.1
6 Guatemala (1976)	7.9
7 Tangshan, China (1976)	7.6
8 Messina, Italy (1908)	7.5
9 Vrancea, Romania (1977)	7.2
10 Agadir, Morocco (1960)	5.9

Richter scale *below*
This table gives joule equivalents for units of the Richter scale. This scale is used to record an earthquake's "magnitude." It is a measure, based on seismograph readings and mathematical formulae, of the total energy released.

Mercalli equivalents *below* The Mercalli scale measures earthquake intensity. Numbers refer to an earthquake's effects at some specific place on Earth's surface. Mercalli numbers and characteristics are listed here with Richter, joule and TNT equivalents.

Earthquake magnitudes *above* Shown on the diagram and listed in the table beside it are the magnitudes of selected earthquakes in different parts of the world. We give the year of the earthquake, and then its magnitude on the Richter scale.

Frequency of earthquakes Each year there are more than 300,000 earth tremors with Richter magnitudes of 2–2.9. An earthquake with a Richter magnitude of 8.5 or over occurs about every 5–10 years.

Richter	Joules
0	6.3×10^{-2}J
1	1.6×10J
2	4.0×10^3J
3	1.0×10^6J
4	2.5×10^8J
5	6.3×10^{10}J
6	1.6×10^{13}J
7	4.0×10^{15}J
8	1.0×10^{18}J
9	2.5×10^{20}J
10	6.3×10^{22}J

Mercalli number and characteristics		Richter	Joules	TNT equivalent
I	Instrumental: detected only by seismographs	<3.5	$<1.6 \times 10^7$J	<7.6 lb
II	Feeble: noticed only by some people at rest	3.5	1.6×10^7J	7.6 lb
III	Slight: similar to vibrations from a passing truck	4.2	7.5×10^8J	357 lb
IV	Moderate: felt generally indoors; parked cars rock	4.5	4.0×10^9J	1905 lb
V	Rather strong: felt generally; most sleepers wake	4.8	2.1×10^{10}J	5t
VI	Strong: trees sway; furniture moves; some damage	5.4	5.7×10^{11}J	136t
VII	Very strong: general alarm; walls crack	6.1	2.8×10^{13}J	6.6kt
VIII	Destructive: weak structures damaged; walls fall	6.5	2.5×10^{14}J	60kt
IX	Ruinous: some houses collapse as ground cracks	6.9	2.3×10^{15}J	550kt
X	Disastrous: many buildings destroyed; rails bend	7.3	2.1×10^{16}J	5Mt
XI	Very disastrous: few buildings survive; landslides	8.1	1.7×10^{18}J	405Mt
XII	Catastrophic: total destruction; ground forms waves	>8.1	$>1.7 \times 10^{18}$J	>405Mt

© DIAGRAM

SOUND

The term "sound" is used to refer both to a physical phenomenon (vibrations transmitted as compression waves through air or some other medium) and to a sensation in our minds (our brain's response to these vibrations after they reach our eardrums). Here we compare the physical characteristics of some different sounds.

Wavelength and frequency
right Sound waves are longitudinal pressure waves. Wavelength is the distance between successive peaks (a). Frequency is the number of waves that pass a given point in a given time; it is calculated by dividing wave speed by wavelength, and is measured in cycles per second or hertz (Hz). The shorter the length of a wave, the higher is its frequency. The longer the wave, the lower is the frequency. High-pitched tones are produced by high-frequency waves; low tones by low-frequency waves.

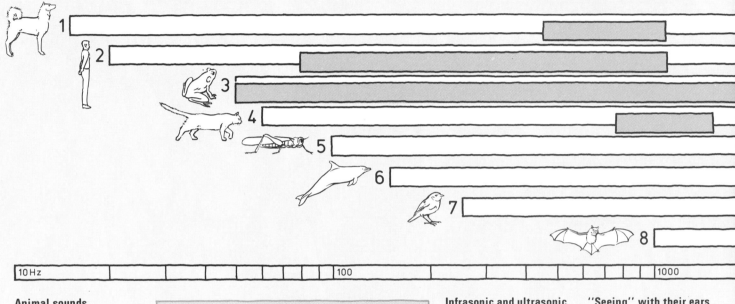

| 10Hz | | | | | | 100 | | | | | 1000 |

Animal sounds
Shown against a logarithmic hertz scale in the diagram *above* and listed in the table *right* are the sound frequency ranges that are heard (white bars) and produced (colored bars) by selected animals and man.

Animal	Range heard	Range produced
1 Dog	15–50,000Hz	452–1080Hz
2 Man	20–20,000Hz	80–1100Hz
3 Frog	50–10,000Hz	50–8000Hz
4 Cat	60–65,000Hz	760–1520Hz
5 Grasshopper	100–15,000Hz	7000–100,000Hz
6 Dolphin	150–150,000Hz	7000–120,000Hz
7 European robin	250–21,000Hz	2000–13,000Hz
8 Bat	1000–120,000Hz	10,000–120,000Hz

Infrasonic and ultrasonic
left, above The normal human ear is sensitive to sounds with frequencies between about 20Hz and 20,000Hz. Sounds with frequencies below 20Hz are described as "infrasonic," those above 20,000Hz as "ultrasonic."

"Seeing" with their ears
Whereas most animals locate objects visually, by seeing reflected light waves, a few animals, notably bats and dolphins, obtain such information via their ears, by picking up the reflected sound waves of their own high-pitched squeaks.

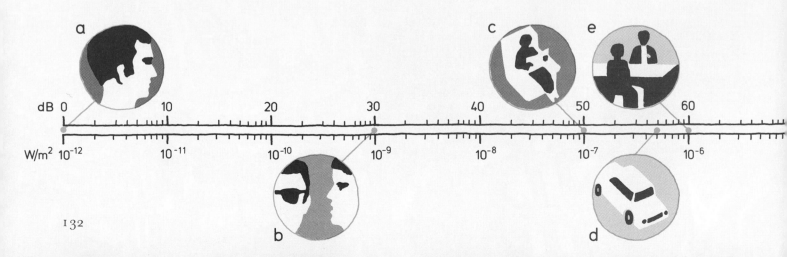

Most of the sounds made by bats are too high to be heard by humans, but the intensity of their squeaks is comparable with the sound of a four-engined jet aircraft only 1 mile away.

Energy

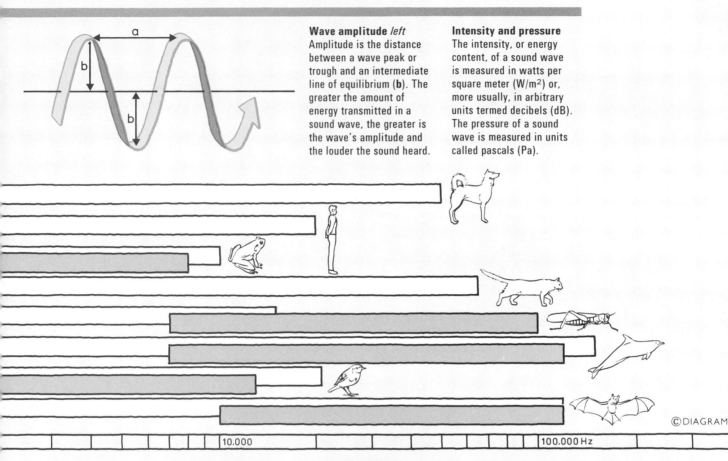

Wave amplitude *left*
Amplitude is the distance between a wave peak or trough and an intermediate line of equilibrium (**b**). The greater the amount of energy transmitted in a sound wave, the greater is the wave's amplitude and the louder the sound heard.

Intensity and pressure
The intensity, or energy content, of a sound wave is measured in watts per square meter (W/m^2) or, more usually, in arbitrary units termed decibels (dB). The pressure of a sound wave is measured in units called pascals (Pa).

©DIAGRAM

A world of noise
The diagram *below* and the table *right* show the relative intensities, in decibels (dB) and in Watts per square meter (W/m^2), of a range of familiar sounds. The minute energy content in any sound wave and the wide range of possible intensities make W/m^2 rather cumbersome units for sound. Decibels are easier to use, and correspond to our perception of "loudness." Starting with 0dB equal to an arbitrary 10^{-12}W/m^2, we have an increase of 10dB of loudness for every tenfold increase in W/m^2 of power.

a Human minimum audibility level	0dB	10^{-12}W/m^2
b Soft whisper at 5m	30dB	10^{-9}W/m^2
c Interior of typical urban home	50dB	10^{-7}W/m^2
d Light traffic at 15m	55dB	5 x 10^{-7}W/m^2
e Average conversation at 1m	60dB	10^{-6}W/m^2
f Pneumatic drill at 15m	85dB	5 x 10^{-4}W/m^2
g Heavy traffic at 15m	90dB	10^{-3}W/m^2
h Loud shout at 15m	100dB	10^{-2}W/m^2
i Jet aircraft take-off at 600m	105dB	5 x 10^{-2}W/m^2
j Discotheque at full volume	117dB	7 x 10^{-1}W/m^2
k Jet aircraft take-off at 60m	120dB	1W/m^2
l Painful level for humans	130dB	10W/m^2
m Jet aircraft take-off at 30m	140dB	10^{2}W/m^2

An ear for trouble *left*
The quietest sounds heard by a person with normal hearing in two ears have an intensity of 10^{-12}W/m^2 (0dB). Such a person experiences pain from sounds of 1–10W/m^2 (120–130dB), and is liable to permanent ear damage if sounds exceed 100W/m^2 (140dB).

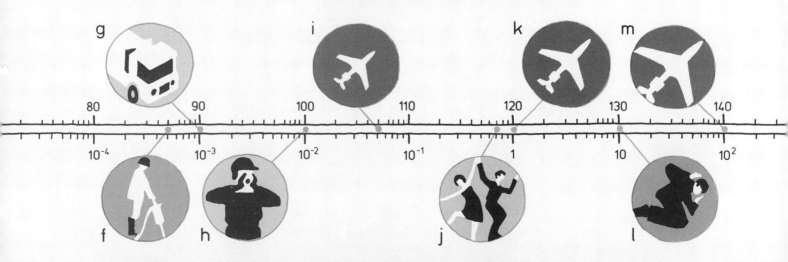

MUSIC

In the sound phenomenon known as music, the most precisely controlled aspect is the pitch or frequency. Instruments are designed, and musicians are trained, to produce frequencies that conform to a basic set. This basic set is most easily seen in the piano, whose range of $7\frac{1}{4}$ octaves is divided into 88 equal steps.

Orchestral instruments
Listed *right* and shown on the diagram *below* are the normal ranges of the piano and some common orchestral instruments. The vertical black line at 440Hz indicates "concert A" (A_4), used as a tuning guide.

All-encompassing piano
The range of the piano encompasses the ranges of all Western orchestral instruments. The piano keys in our diagram correspond as closely as possible to the hertz scale labeled beneath them, and help to illustrate the ranges of tones produced by other instruments.

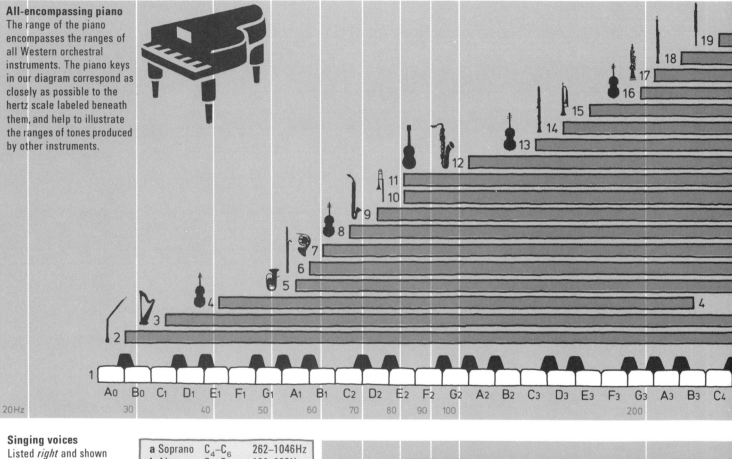

Singing voices
Listed *right* and shown *far right* against the same scale as our musical instrument ranges are the normal ranges for trained singing voices, from the bass's lowest note of E_2 (82.4Hz) to the soprano's highest C_6 (1046Hz).

a Soprano	C_4–C_6	262–1046Hz
b Alto	G_3–F_5	196–698Hz
c Tenor	D_3–Bb_4	147–466Hz
d Baritone	A_2–G_4	110–392Hz
e Bass	E_2–D_4	82.4–294Hz

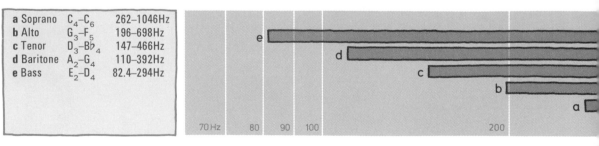

The normal range of the tenor trombone extends from the lowest note of a bass singer's normal range to the highest note for a tenor singer.

For tuning purposes the pitch of the A above middle C was in 1939 internationally agreed at 440Hz. Pitches formerly accepted for this same A include: 435Hz (Paris Academy 1859 and the Vienna conference of 1885); 415–425Hz (representative pitch standards in the early 18th century); 377Hz (Arnold Schlick "low organ" pitch of 1511).

1	Piano	A_0–C_8	27.5–4186Hz
2	Contrabassoon	Bb_0–Eb_4	29–311Hz
3	Harp	C_1–G_7	32.7–3136Hz
4	Double bass	E_1–B_3	41.2–247Hz
5	Eb bass tuba	A_1–Eb_4	55–311Hz
6	Bassoon	Bb_1–Eb_5	58.3–622Hz
7	French horn	B_1–F_5	61.7–698Hz
8	Violoncello	C_2–E_5	65.4–659Hz
9	Bass clarinet	D_2–Eb_5	73.4–622Hz
10	Tenor trombone	E_2–Bb_4	82.4–466Hz
11	Guitar	E_2–F_5	82.4–698Hz
12	Tenor saxophone	Ab_2–Eb_5	104–622Hz
13	Viola	C_3–C_6	131–1046Hz
14	Clarinet	D_3–G_6	147–1568Hz
15	Trumpet	E_3–Bb_5	165–932Hz
16	Violin	G_3–C_7	196–2093Hz
17	Soprano saxophone	Ab_3–Eb_6	208–1244Hz
18	Oboe	Bb_3–F_6	233–1397Hz
19	Flute	C_4–C_7	262–2093Hz
20	Piccolo	D_5–Bb_7	587–3729Hz

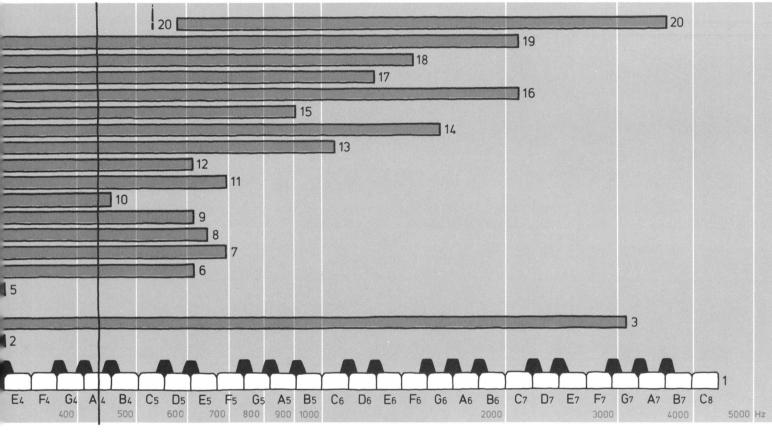

Heights and depths *right* Shown here are musicians playing the lowest and the highest sounding instruments of the orchestral woodwind and string families.

A Contrabassoon 29–311Hz
B Double bass 41.2–247Hz
C Violin 196–2093Hz
D Piccolo 587–3729Hz

©DIAGRAM

ELECTROMAGNETIC WAVES

The energy of the Sun and of the local radio station are both transmitted by electromagnetic waves. Here we compare the characteristics of waves within different regions of the electromagnetic spectrum, and look at some of the practical applications that man has found for them in laboratories, homes and hospitals.

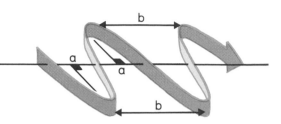

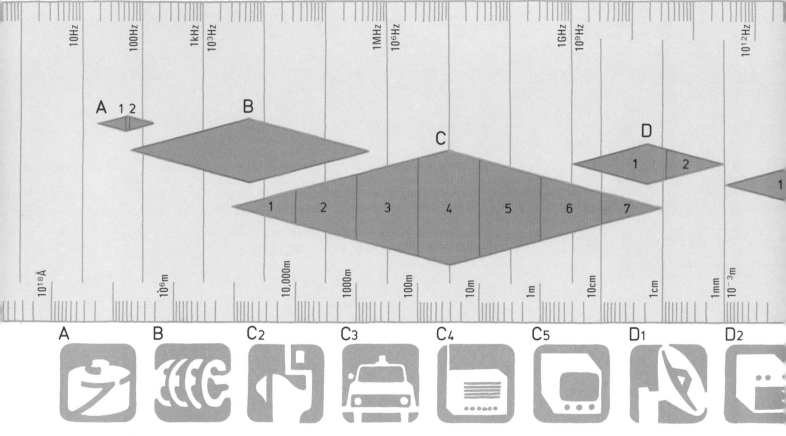

Electromagnetic spectrum
above The top logarithmic scale gives frequencies in hertz; the lower one gives equivalent wavelength measurements. The lower the frequency the longer is the wavelength; the higher the frequency the shorter the wavelength will be. Hertz measurements are used for low frequency waves, and wavelength measurements for higher frequency waves. Sections of the spectrum, described *right*, are located between the scales, and illustrated with examples indicating their use or character.

A) Generated electricity
The alternating current (AC) electricity used in some countries for domestic and commercial use is at the low frequency/long wavelength end of the spectrum, and ranges from 16.7–144Hz. Two lines (**1,2**) on diamond **A** indicate two common frequencies: 50Hz (Europe) and 60Hz (North America). These are heard as the "mains hum" from faulty domestic gadgetry, hi-fi systems, etc.

B) Induction heating
Waves in this frequency band are used to induce heat in metals. The metal is placed in the center of a series of wire coils through which current is passed, inducing electrical "eddy currents" in the metal, which raise its temperature. Commonly the frequency of the current supplied to the coil is in the range of 60–60,000Hz, but waves with frequencies up to 500,000Hz are used.

C) Radio waves
These are the waves used for radio and television transmission. They range from 3kHz to 30GHz and, as shown on diamond **C**, may be subdivided as follows.
1 Very low frequency (vlf).
2 Low frequency (lf) waves, used for ship radio signals.
3 Medium frequency (mf), as used by police forces.
4 High frequency (hf), used for "shortwave" radio.
5 Very high frequency (vhf), used for radio and for television.
6 Ultra high frequency (uhf).
7 Super high frequency (shf).

D) Microwaves
Waves in this region are used in radar (**ra**dio **d**etecting **an**d **r**anging), a method of detecting otherwise invisible objects by bouncing radio pulses off them. Such pulses are transmitted along a carrier wave in the range 1–35GHz (diamond **D**, region 1). The shorter wavelengths in the microwave region, down to 1mm (diamond **D**, region 2), have found application in cooking, greatly reducing cooking times.

F

Tests indicate that, according to average judgment, names for the pure spectral colors should be applied to light of the following wavelengths:

Color	Wavelength
Violet	3900–4550Å
Blue	4550–4920Å
Green	4920–5770Å
Yellow	5770–5970Å
Orange	5970–6220Å
Red	6220–7700Å

Energy

Wave characteristics

Electromagnetic waves are produced by rhythmic variations in electrical and magnetic fields. As shown in the diagram *left* they are "transverse" waves, vibrating in a plane at right angles (**a**) to the direction of flow.

Measurement of waves

Electromagnetic waves are measured in two ways: by "wavelength" (the distance from one wave peak to another, **b** on the diagram *left*) and by "frequency" (the number of waves, or "cycles," per second). The table *right* gives units of measurement and methods of converting frequency measurements to wavelength measurements, and vice versa. All waves within the electromagnetic spectrum travel, or "propagate themselves," at a constant speed, the speed of light (approximately 3×10^8m/s).

Units of frequency
1000 hertz (Hz) = 1 kilohertz (kHz)
1000 kilohertz = 1 megahertz (MHz)
1000 megahertz = 1 gigahertz (GHz)
Units of wavelength
1000 X-units (Xu) = 1 Ångstrom (Å) = 10^{-10}m
10,000 Ångstroms = 1 micron (μ) = 10^{-6}m
1000 microns = 1 millimeter = 10^{-3}m
Conversions
Wavelength (meters) = 3×10^8m/s ÷ frequency (hertz)
Frequency (hertz) = 3×10^8m/s ÷ wavelength (meters)

©DIAGRAM

E F G H I

2 3   2   1 2

10^{15}Hz 10^{18}Hz 10^{21}Hz 10^{24}Hz

1μ 10^{-6}m 1000Å 100Å 10Å 10^{-9}m 1Å 100Xu 10Xu 10^{-12}m 1Xu 10^{-15}m 10^{-6}Å

E G H₁ H₂ I

E) Infrared waves

These are heat waves, with wavelengths between about 1mm and 7700Å. They are subdivided into far, middle, and near infrared (**1, 2, 3** on diamond **E**). Most of the energy radiated by hot objects (including the energy we receive from the Sun) lies chiefly in this range. Practical applications are in the areas of photography (plates that are sensitive to infrared radiation make it possible to take photographs in the dark), physical therapy, military reconnaissance, and astronomical research.

F) Light waves

Although scientists use it more widely, the term "light" is most commonly taken to refer to the visible portion of the electromagnetic spectrum, ie to wavelengths of approximately 3900–7700Å. The color of an object is determined by the composition of its surface, which reflects certain wavelengths but not others. White objects reflect all wavelengths of the visible part of the spectrum; black objects reflect none of them.

G) Ultraviolet light

This highly energetic radiation is invisible to the human eye, having wavelengths of about 3900–100Å. As indicated on diamond **G** it is subdivided into near (**1**) and far (**2**) regions. Earth's atmosphere absorbs much of this type of radiation from the Sun. Ultraviolet light causes sunburn, and helps in the formation of vitamin D within the body. It also causes some substances to fluoresce (invisible waves are absorbed and visible ones emitted).

H) X-rays

These are even more energetic than ultraviolet light, with wavelengths of approximately 100Å to 30Xu. They have found wide application because of their ability to penetrate many substances. The science of X-ray spectroscopy uses the lower end of the X-ray frequency range (**1**). Higher frequency uses include the photographing of bones and the detection of flaws in metal objects (**2**).

I) Gamma rays

These are produced when atomic nuclei pass from a higher to a lower energy level. They are received from distant nuclear explosions and from nuclear reactions in the universe, and are also produced by the interaction of so-called cosmic rays with particles in the atmosphere. Gamma rays have very high energy levels (10^4–10^7eV), and can pass through several centimeters of lead. The lowest frequencies of gamma rays have wavelengths of about 1Xu; in theory there is no top limit.

FOOD

People, like other animals, obtain their energy from the food they eat. Here we compare the energy contents of different foods—from butter, with 1625 Calories per ½lb, to celery, with only 7 Calories in an 8 inch outer stalk. On the following pages we compare Calorie requirements for different people and for various activities.

Calories and calories
The energy content of food is measured in Calories (written with a capital C). One Calorie is equivalent to one kilocalorie and to 1000 calories (small c). For fuller definitions and for conversions into other energy units, see p. 126.

A full table of food
right Included here is a selection of different types of food, with the number of Calories contained in specified quantities of each. As far as possible, the quantities specified are for an "average" adult portion.

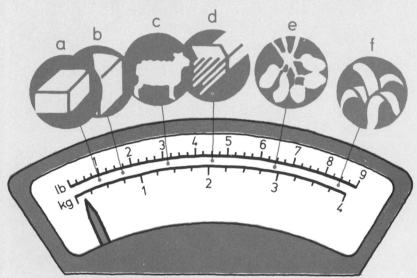

Calorie food scale
The diagram *above* and the list *right* show how much a young man would have to eat if he were to obtain his required daily intake of 3000 Calories (see p. 140) from eating only one type of food.

a Butter 15oz
b Cheddar cheese 1 lb 11oz
c Shoulder of lamb, roasted lean meat 3 lb 3oz
d Spaghetti, boiled 4 lb 9oz
e Potatoes, boiled or baked 6 lb 8oz
f Bananas 8 lb 8oz

Only 15oz of butter contains enough Calories to meet a young man's daily requirement of 3000 Calories. If he were to obtain the same number of Calories from lettuce, it would be necessary to eat 37lb 8oz of it.

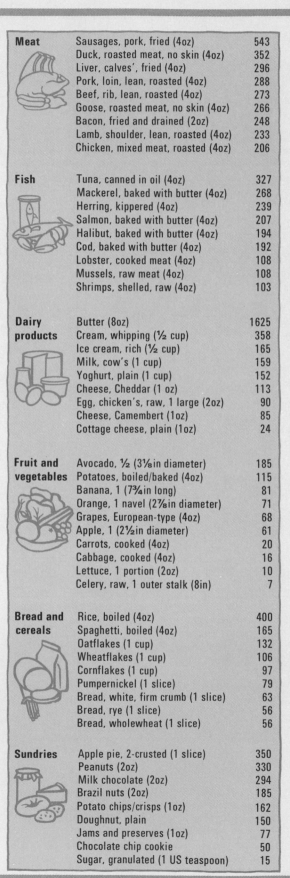

Meat		
	Sausages, pork, fried (4oz)	543
	Duck, roasted meat, no skin (4oz)	352
	Liver, calves', fried (4oz)	296
	Pork, loin, lean, roasted (4oz)	288
	Beef, rib, lean, roasted (4oz)	273
	Goose, roasted meat, no skin (4oz)	266
	Bacon, fried and drained (2oz)	248
	Lamb, shoulder, lean, roasted (4oz)	233
	Chicken, mixed meat, roasted (4oz)	206
Fish	Tuna, canned in oil (4oz)	327
	Mackerel, baked with butter (4oz)	268
	Herring, kippered (4oz)	239
	Salmon, baked with butter (4oz)	207
	Halibut, baked with butter (4oz)	194
	Cod, baked with butter (4oz)	192
	Lobster, cooked meat (4oz)	108
	Mussels, raw meat (4oz)	108
	Shrimps, shelled, raw (4oz)	103
Dairy products	Butter (8oz)	1625
	Cream, whipping (½ cup)	358
	Ice cream, rich (½ cup)	165
	Milk, cow's (1 cup)	159
	Yoghurt, plain (1 cup)	152
	Cheese, Cheddar (1 oz)	113
	Egg, chicken's, raw, 1 large (2oz)	90
	Cheese, Camembert (1oz)	85
	Cottage cheese, plain (1oz)	24
Fruit and vegetables	Avocado, ½ (3⅛in diameter)	185
	Potatoes, boiled/baked (4oz)	115
	Banana, 1 (7¾in long)	81
	Orange, 1 navel (2⅞in diameter)	71
	Grapes, European-type (4oz)	68
	Apple, 1 (2½in diameter)	61
	Carrots, cooked (4oz)	20
	Cabbage, cooked (4oz)	16
	Lettuce, 1 portion (2oz)	10
	Celery, raw, 1 outer stalk (8in)	7
Bread and cereals	Rice, boiled (4oz)	400
	Spaghetti, boiled (4oz)	165
	Oatflakes (1 cup)	132
	Wheatflakes (1 cup)	106
	Cornflakes (1 cup)	97
	Pumpernickel (1 slice)	79
	Bread, white, firm crumb (1 slice)	63
	Bread, rye (1 slice)	56
	Bread, wholewheat (1 slice)	56
Sundries	Apple pie, 2-crusted (1 slice)	350
	Peanuts (2oz)	330
	Milk chocolate (2oz)	294
	Brazil nuts (2oz)	185
	Potato chips/crisps (1oz)	162
	Doughnut, plain	150
	Jams and preserves (1oz)	77
	Chocolate chip cookie	50
	Sugar, granulated (1 US teaspoon)	15

Calories

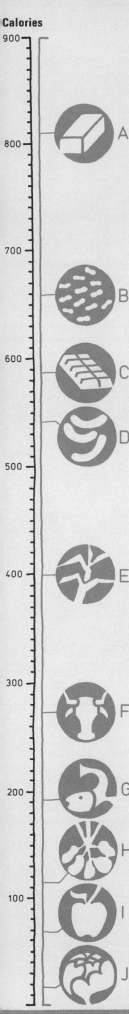

A large egg contains 90 Calories
if it is served boiled or poached,
108 Calories if it is fried, and
120 Calories if it is scrambled.

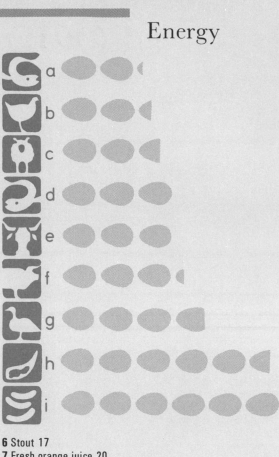

Calories by the quarter
left This diagram gives
a visual comparison of
the number of Calories
contained in 4oz of a
selection of foods from the
table on the previous page.
A Butter 812.5
B Peanuts 660
C Milk chocolate 588
D Pork sausages (fried) 543
E Rice (boiled) 400
F Beef (roast, lean rib) 273
G Cod (buttered, baked) 192
H Potatoes (boiled) 115
I Apple 68
J Cabbage (boiled) 16

Energy in eggs
right Here we express the
Calorie contents of 4oz of
different types of meat and
fish in terms of the number
of large boiled eggs that
would have to be eaten to
provide the same number of
Calories. (Cuts and cooking
methods are the same as
those in the main table.)
a Cod 2.1
b Chicken 2.3
c Lamb 2.6
d Mackerel 3.0
e Beef 3.0
f Pork 3.2
g Duck 3.9
h Bacon 5.5
i Pork sausages 6

A drink or two or more
The diagram *below* and the
list *right* provide a
comparison of the Calorie
contents of various drinks.
Shown are the number of
glasses (4 US fl.oz) of each
drink that would have to be
drunk to provide 1200
Calories, the number of
Calories in a US pint bottle
of 90° proof distilled spirits.

1 Southern Comfort 2.5
2 Distilled spirits (90°) 4
3 Sweet vermouth 6.7
4 Beaujolais 12.5
5 Reisling wine 13.3

6 Stout 17
7 Fresh orange juice 20
8 Lager 23.1
9 Cola 25
10 Tonic water 31.6

One 1oz glass of whisky together with 2oz
of peanuts has the same number of Calories
(405) as a meal consisting of 4oz of chicken,
a 4oz baked potato, 4oz of cooked cabbage,
and 4oz of grapes.

FOOD AND WORK

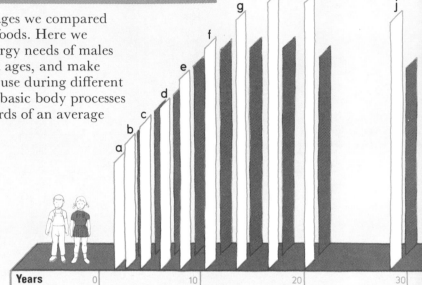

On the previous two pages we compared the energy contents of foods. Here we compare daily food energy needs of males and females of different ages, and make comparisons of Calorie use during different activities. Maintaining basic body processes burns up about two thirds of an average adult's Calorie intake.

Calories per day

The diagram *right* and the table *far right* show estimated daily Calorie (kilocalorie) needs for males and females of different ages. Sex and age are the most important factors affecting Calorie needs, but size, physical activity, and climate are also important. Male and female needs are similar until puberty, thereafter males typically need more Calories than females of similar age. Children use up a lot of Calories for their size, because energy is needed for growth.

Activity	Calories used per hour	
	Males	**Females**
A Sleeping	65	55
B Sitting	90	70
C Standing	120	100
D Walking	220	180
E Walking uphill	440	360
F Running	600	420

Use of Calories

The table *above* and the diagram *right* give estimates of the number of Calories needed by average men and women to perform particular activities for one hour. Obviously the amount of effort any individual puts into an activity affects results, but a general pattern of Calorie use is evident. Men use more Calories than women for all activities, because men have more weight to carry around and because women usually have more body fat and so need less energy to retain body heat.

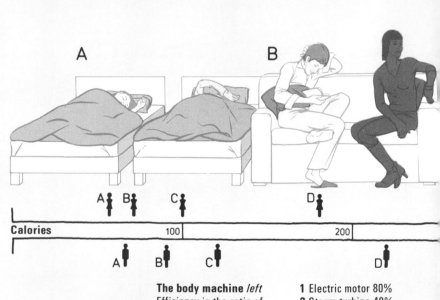

The body machine *left*
Efficiency is the ratio of output (work) to input (fuel/food). Here we compare the efficiency of the human body with that of four other machines. An unfit person's normal 16–27% can be raised to 56% with training.

1 Electric motor 80%
2 Steam turbine 40%
3 Petrol motor 20–30%
4 Steam engine 10–15%
5 Human body 16–56%

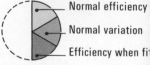

— Normal efficiency
— Normal variation
— Efficiency when fi

An 18-year-old male's 3000 Calories a day is equivalent to the energy contained in 13oz of good quality bituminous coal. An 18-year-old female's 2300 Calories is equivalent to 10oz of the same grade of coal.

A grandfather aged 75 requires the same number of Calories per day as his grandson aged 8. A grandmother aged 75 has the same daily Calorie needs as her 6-year-old granddaughter.

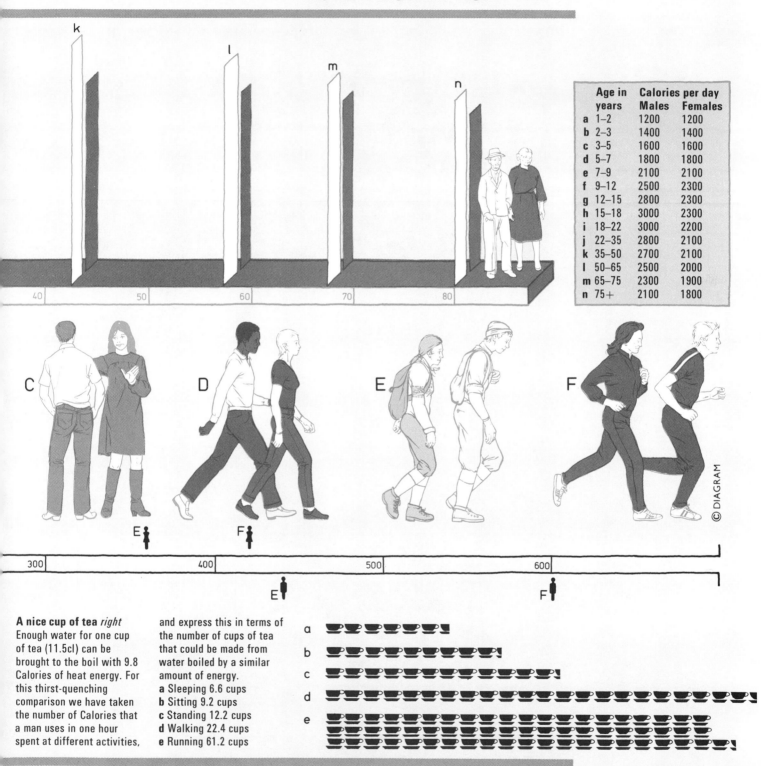

	Age in years	Calories per day Males	Females
a	1–2	1200	1200
b	2–3	1400	1400
c	3–5	1600	1600
d	5–7	1800	1800
e	7–9	2100	2100
f	9–12	2500	2300
g	12–15	2800	2300
h	15–18	3000	2300
i	18–22	3000	2200
j	22–35	2800	2100
k	35–50	2700	2100
l	50–65	2500	2000
m	65–75	2300	1900
n	75+	2100	1800

© DIAGRAM

A nice cup of tea *right*
Enough water for one cup of tea (11.5cl) can be brought to the boil with 9.8 Calories of heat energy. For this thirst-quenching comparison we have taken the number of Calories that a man uses in one hour spent at different activities, and express this in terms of the number of cups of tea that could be made from water boiled by a similar amount of energy.
a Sleeping 6.6 cups
b Sitting 9.2 cups
c Standing 12.2 cups
d Walking 22.4 cups
e Running 61.2 cups

During 8 hours of sleep a man uses the same number of Calories as if he had walked for 2 hours and 22 minutes or run for 52 minutes.

CHAPTER 7

LES GRANDES DUNES D'EL BAB.

Photograph from a book published in 1912 showing a nomad and his camel traversing the Libyan Sahara, where the world's highest air temperatures are recorded.

Mount Erebus—a 12,200ft active volcano—towers above the icy wastes of Antarctica in this painting by Edward Wilson, who went with Scott in 1901–04 (Abbot Hall Art Gallery, Kendal, England).

TEMPERATURE

An engraving showing an experiment by the 18th-century French scientist Antoine-Laurent Lavoisier in which he used two large lenses to focus the heat energy of the Sun in order to ignite alcohol in a container (Science Museum, London).

MEASURING TEMPERATURE

Temperature is a measurement of heat energy made on one of the scales devised for this purpose. Fahrenheit, Réaumur and Celsius based their scales on the boiling and freezing points of water. The scales of Kelvin and Rankine were based on "absolute zero," the theoretical point at which there is no temperature or pressure.

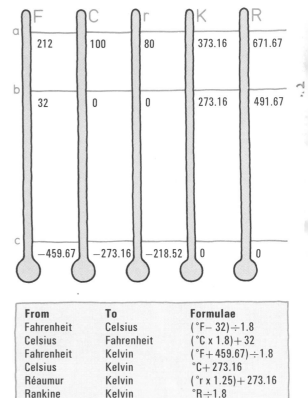

	F	C	r	K	R
a	212	100	80	373.16	671.67
b	32	0	0	273.16	491.67
c	−459.67	−273.16	−218.52	0	0

High tea? The diagram *left* and the table *below* show the boiling point of water at different altitudes. Differences occur because atmospheric pressure falls as altitude increases, and the lower the pressure, the lower the boiling point of water (and vice versa).

Temperature scales *right* This diagram shows the values of the boiling (**a**) and freezing (**b**) points of water and of absolute zero (**c**) on five temperature scales. Below this are the formulae for converting one to another. The Fahrenheit, Celsius and Réaumur (now obsolete) scales divide the distance between the melting and boiling points of water arbitrarily into 180°F, 100°C and 80°r. The Kelvin and Rankine scales begin at "absolute zero." The Kelvin unit (K) equals the Celsius degree, and the Rankine degree (°R) equals the Fahrenheit degree.

Place	Altitude	Water boils	
A London, England	Sea level	212.0°F	100°C
B Dead Sea	−1296ft	213.8°F	101°C
C Denver, Colorado	5280ft	203.0°F	95°C
D Quito, Ecuador	9350ft	194.0°F	90°C
E Lhasa, Tibet	12,087ft	188.6°F	87°C
F Mt Everest (top)	29,002ft	159.8°F	71°C

From	To	Formulae
Fahrenheit	Celsius	(°F− 32)÷1.8
Celsius	Fahrenheit	(°C x 1.8)+ 32
Fahrenheit	Kelvin	(°F+ 459.67)÷1.8
Celsius	Kelvin	°C+ 273.16
Réaumur	Kelvin	(°r x 1.25)+ 273.16
Rankine	Kelvin	°R÷1.8
Kelvin	Fahrenheit	(K x 1.8) − 459.67
Kelvin	Celsius	K− 273.16

Universal temperatures *below* Scale **A**, extending the full width of these pages, is a logarithmic Kelvin temperature scale. It runs, left to right, from very high temperatures (10^{10}K) to very low ones (below 0.0001K). Drawn in their equivalent positions alongside the Kelvin scale are a Celsius (also called Centigrade) scale (**B**) and a Fahrenheit scale (**C**). The Kelvin scale measures absolute temperatures and because of this makes an ideal universal scale; the Celsius and Fahrenheit scales, being based on the boiling and freezing points of water, are rarely used to express the greatest extremes.

Very high temperatures Selected high-temperature phenomena are included in the list *right*—with temperatures expressed in Kelvins. The same examples have also been plotted on the universal temperature scale *below*, and then illustrated beneath it.

1 Supernova 3.5 x 10^9K
2 Interior of hottest stars, in excess of 10^9K
3 Thermonuclear explosion 10^8K
4 Sun's interior 2 x 10^7K
5 Sun's corona 10^6K
6 Molecules break down into atoms 5000K
7 Temperature of a lamp's tungsten filament 4000K
8 Heat of domestic coal gas flame 2500K
9 Temperature of molten lava 2000K

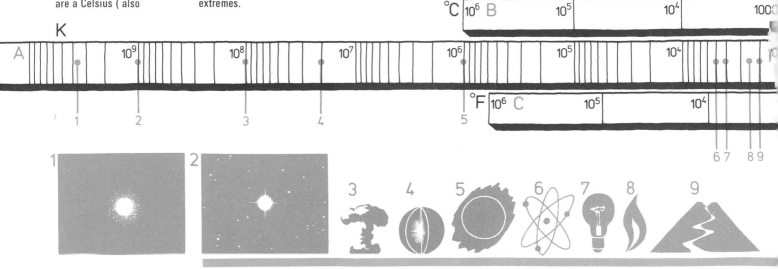

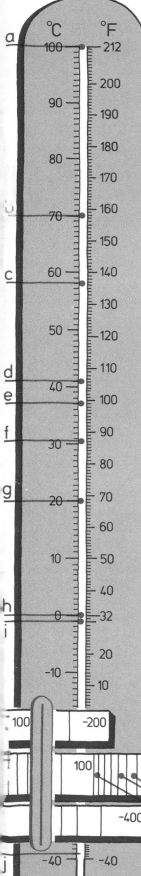

Temperature

The highest man-made temperatures, at the center of thermonuclear bombs, are in the order of $3 \times 10^8 K$. The lowest temperature ever reached in experiments is $5 \times 10^{-7} K$.

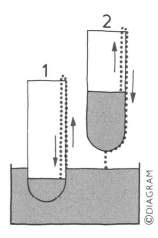
©DIAGRAM

Fahrenheit and Celsius
right In the first two pairs of columns we give the Celsius equivalents for every five degrees Fahrenheit from —40°F to 220°F. In the third pair of columns we give the Fahrenheit equivalents for every five Celsius degrees from —40°C to 100°C.
One Fahrenheit degree is equivalent to 0.5 recurring (taken as 0.56) of a Celsius degree. One Celsius degree is equivalent to 1.8 Fahrenheit degrees. To find the equivalents for the figures between those listed, add or subtract the appropriate amount from the nearest figure given.

°F	°C	°F	°C	°C	°F
—40 =	—40.00	100 =	37.78	—40 =	—40
—35 =	—37.23	105 =	40.56	—35 =	—31
—30 =	—34.45	110 =	43.34	—30 =	—22
—25 =	—31.67	115 =	46.11	—25 =	—13
—20 =	—28.89	120 =	48.89	—20 =	—4
—15 =	—26.12	125 =	51.66	—15 =	5
—10 =	—23.34	130 =	54.45	—10 =	14
—5 =	—20.56	135 =	57.23	—5 =	23
0 =	—17.78	140 =	60.00	0 =	32
5 =	—15.00	145 =	62.78	5 =	41
10 =	—12.23	150 =	65.56	10 =	50
15 =	9.45	155 =	68.34	15 =	59
20 =	—6.67	160 =	71.12	20 =	68
25 =	—3.89	165 =	73.89	25 =	77
30 =	—1.11	170 =	76.67	30 =	86
35 =	1.67	175 =	79.45	35 =	95
40 =	4.45	180 =	82.23	40 =	104
45 =	7.23	185 =	85.00	45 =	113
50 =	10.00	190 =	87.78	50 =	122
55 =	12.78	195 =	90.56	55 =	131
60 =	15.56	200 =	93.34	60 =	140
65 =	18.34	205 =	96.11	65 =	149
70 =	21.11	210 =	98.89	70 =	158
75 =	23.89	215 =	101.67	75 =	167
80 =	26.67	220 =	104.44	80 =	176
85 =	29.45			85 =	185
90 =	32.23			90 =	194
95 =	35.00			95 =	203
				100 =	212

Superfluid helium *above*
Helium liquefies at 4.2K; at 2.2K it becomes "superfluid." If an empty tube is lowered into superfluid helium the helium creeps up the sides of the tube (**1**) to fill it to the level of the helium outside. If the tube is then removed from the liquid in the container, the helium in the tube climbs back out and down the sides of the tube (**2**)!

Everyday temperatures
The thermometer *left* is an enlargement of that over the main scale *below* and shows the temperature range with which we are most familiar. Examples falling within this range are plotted on the thermometer, showing °C and °F, and listed *right*.

a Water boils 212°F/100°C (373K)
b Some bacteria survive 158°F/70°C (343K)
c Highest recorded shade temperature 136.4°F/58°C (331K)
d Sparrow's mean body temperature 106°F/41°C (314K)
e Normal body temperature for humans 98.6°F/37°C (310K)
f Butter melts 87°F/30.6°C (304K)
g Comfortable room temperature 68°F/20°C (293K)
h Pure water freezes 32°F/0°C (273K)
i Temperature of Arctic seawater 30°F/—1.1°C (272K)
j Mercury freezes —38.0°F/—38.87°C (234.29K)

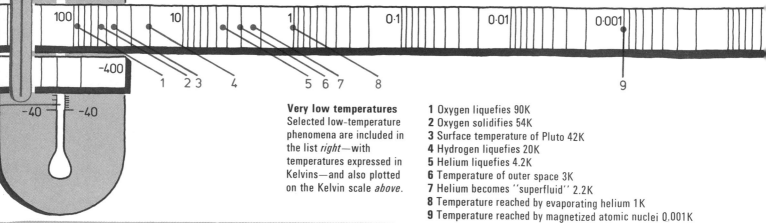

Very low temperatures
Selected low-temperature phenomena are included in the list *right*—with temperatures expressed in Kelvins—and also plotted on the Kelvin scale *above*.

1 Oxygen liquefies 90K
2 Oxygen solidifies 54K
3 Surface temperature of Pluto 42K
4 Hydrogen liquefies 20K
5 Helium liquefies 4.2K
6 Temperature of outer space 3K
7 Helium becomes "superfluid" 2.2K
8 Temperature reached by evaporating helium 1K
9 Temperature reached by magnetized atomic nuclei 0.001K

FREEZING, MELTING AND BOILING POINTS

All matter can exist in three states: as a solid, a liquid or a gas. The particular state of matter at a given time depends on the amount of energy it contains, which is measured by its temperature. Here we compare the freezing/melting and boiling points of selected elements, with one another and with other temperature phenomena.

In different states *below, right* Shown here are the temperatures at which water and selected elements pass from one state to another (see key beneath the first diagram). In the lists beside the diagrams the first figure is the freezing/melting point and the second the boiling point; Celsius temperatures are given in accordance with scientific practice. The range of temperatures through which water is liquid is marked as a band (**A**) in the middle of the first diagram and at the left of the second.

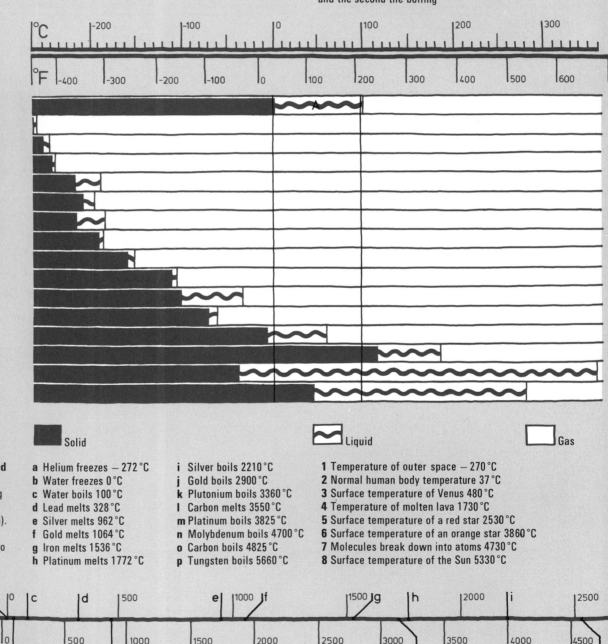

Water 0/100 °C
Helium − 272/− 269 °C
Hydrogen − 259/− 253 °C
Neon − 249/− 246 °C
Fluorine − 220/− 188 °C
Nitrogen − 210/− 196 °C
Oxygen − 219/− 183 °C
Argon − 189/− 186 °C
Krypton − 157/− 152 °C
Xenon − 112/− 107 °C
Chlorine − 101/− 35 °C
Radon − 71/− 62 °C
Bromine − 7/59 °C
Iodine 114/184 °C
Mercury − 39/357 °C
Phosphorus 44/280 °C

■ Solid   〰 Liquid   □ Gas

Temperatures compared
Plotted *below* and listed *right* are freezing/melting and boiling points of elements and water (**a–p**). Other temperature phenomena (**1–8**) are also included for comparison.

a Helium freezes − 272 °C
b Water freezes 0 °C
c Water boils 100 °C
d Lead melts 328 °C
e Silver melts 962 °C
f Gold melts 1064 °C
g Iron melts 1536 °C
h Platinum melts 1772 °C

i Silver boils 2210 °C
j Gold boils 2900 °C
k Plutonium boils 3360 °C
l Carbon melts 3550 °C
m Platinum boils 3825 °C
n Molybdenum boils 4700 °C
o Carbon boils 4825 °C
p Tungsten boils 5660 °C

1 Temperature of outer space − 270 °C
2 Normal human body temperature 37 °C
3 Surface temperature of Venus 480 °C
4 Temperature of molten lava 1730 °C
5 Surface temperature of a red star 2530 °C
6 Surface temperature of an orange star 3860 °C
7 Molecules break down into atoms 4730 °C
8 Surface temperature of the Sun 5330 °C

Mercury, which freezes at −39°C, is unsuitable for thermometers to be used in the coldest regions of the world. Thermometers containing pure alcohol, with a freezing point of −114°C, will record all but the very lowest freak temperatures.

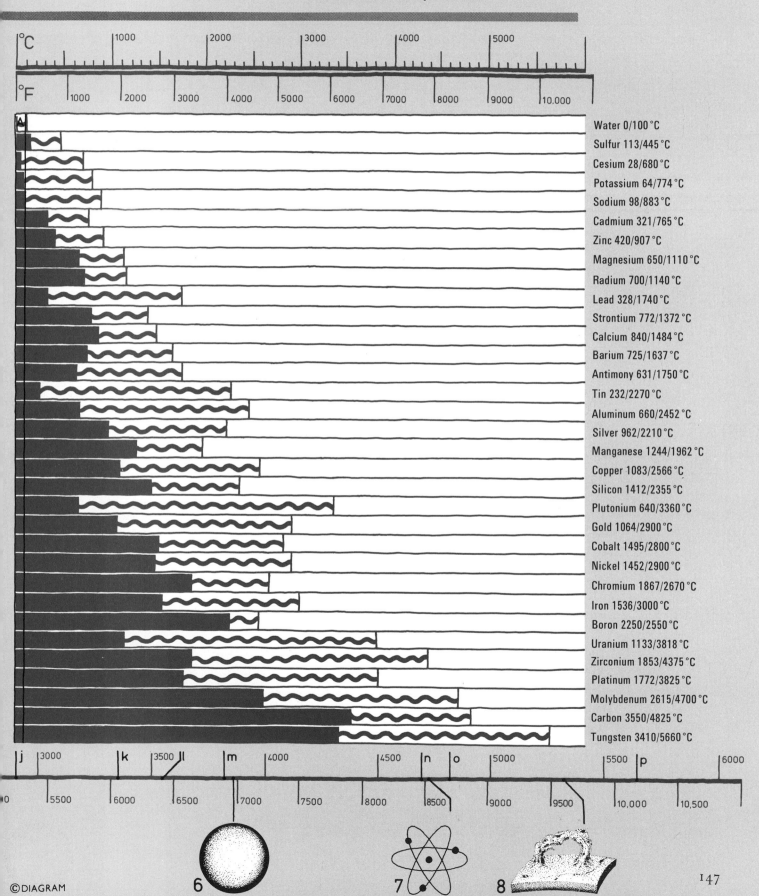

Water 0/100 °C
Sulfur 113/445 °C
Cesium 28/680 °C
Potassium 64/774 °C
Sodium 98/883 °C
Cadmium 321/765 °C
Zinc 420/907 °C
Magnesium 650/1110 °C
Radium 700/1140 °C
Lead 328/1740 °C
Strontium 772/1372 °C
Calcium 840/1484 °C
Barium 725/1637 °C
Antimony 631/1750 °C
Tin 232/2270 °C
Aluminum 660/2452 °C
Silver 962/2210 °C
Manganese 1244/1962 °C
Copper 1083/2566 °C
Silicon 1412/2355 °C
Plutonium 640/3360 °C
Gold 1064/2900 °C
Cobalt 1495/2800 °C
Nickel 1452/2900 °C
Chromium 1867/2670 °C
Iron 1536/3000 °C
Boron 2250/2550 °C
Uranium 1133/3818 °C
Zirconium 1853/4375 °C
Platinum 1772/3825 °C
Molybdenum 2615/4700 °C
Carbon 3550/4825 °C
Tungsten 3410/5660 °C

STARS AND PLANETS

Temperatures in the universe range from an estimated 3,500,000,000K for a supernova to only 3K for the temperature of outer space. Here we look at estimated surface temperatures of stars, the Sun, the planets and of our Moon. Also compared are estimated interior temperatures of the Sun, Earth and Jupiter.

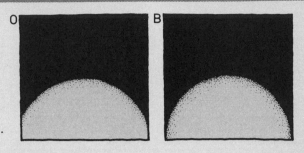

| K | 40,000 | | | | 30,000 | | |
| °F | 70,000 | | 60,000 | | 50,000 | | 40,000 |

Temperatures of stars
Illustrated *top* are examples of different types of star, identified by letters relating to their classification by color. Plotted on the scale *above* and listed in the table *right* are calculated ''surface'' temperatures for typical stars of each class.

There is a clear relationship between a star's color and its surface temperature; blue-white stars are very much hotter than stars with a maximum emission at the orange-red end of the color spectrum. Our Sun is a class G yellow star.

Class	Color	Effective surface temperature	
O	Blue-white	40,000K	71,500°F
B	Blue-white	15,500K	27,400°F
A	White	8500K	14,800°F
F	Yellow-white	6580K	11,380°F
G	Yellow	5520K	9480°F
K	Orange	4130K	6970°F
M	Red	2800K	4580°F

Solar temperatures *left*
This section diagram shows the zones produced as energy travels outward from the Sun's core (**a**). The list beside the diagram identifies the different zones and gives estimated temperatures in Kelvins. The chromosphere (**e**) and the corona (**f**) have higher temperatures than some zones nearer to the Sun's core, a consequence of the frictional action and high velocity of particles within these outer zones. The temperature of the Sun's ''surface'' (within the photosphere) is usually taken as 5600K (9620°F).

a Core 20,000,000K
b Zone of radiation
c Zone of convection
d Photosphere:
inner region 10,000K
outer region 4200K
e Chromosphere:
inner region 4500K
outer region 1,000,000K
f Corona 2,000,000K

The Sun, a class G yellow star, has an estimated surface temperature of 5600K. This makes it twice as hot as a typical class M red star, but only just over one seventh as hot as a typical class O, blue-white star.

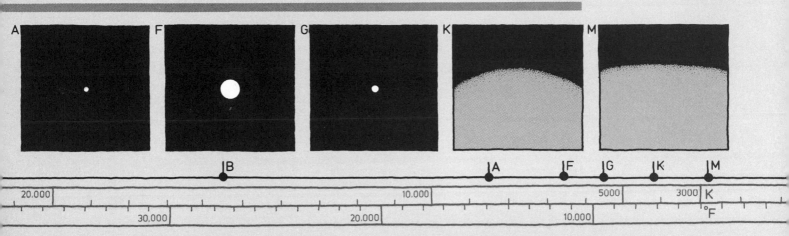

K
°F

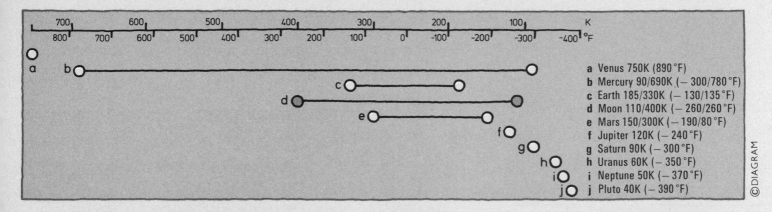

a Venus 750K (890 °F)
b Mercury 90/690K (− 300/780 °F)
c Earth 185/330K (− 130/135 °F)
d Moon 110/400K (− 260/260 °F)
e Mars 150/300K (− 190/80 °F)
f Jupiter 120K (− 240 °F)
g Saturn 90K (− 300 °F)
h Uranus 60K (− 350 °F)
i Neptune 50K (− 370 °F)
j Pluto 40K (− 390 °F)

©DIAGRAM

Hot and cold planets
above Compared here are estimated surface temperatures of the planets and of our Moon. For the four dense inner planets and Moon (**a–e**) we give temperatures for the solid surface. For the four huge low-density outer planets (**f–i**) and for distant Pluto (**j**), the generally accepted cloud "surface" temperatures are given.

Inside stories *right*
Here we compare estimated temperatures for Earth (**1**) and Jupiter (**2**). The core of Jupiter is believed to be over six times hotter than Earth's core, whereas its "surface" is more than twice as cold as Earth's.
1 Earth temperature
a Core 4500K (7600 °F)
b Base of mantle 3300K (5500 °F)
c Surface (average) 295K (70 °F)
2 Jupiter temperature
a Core 30,000K (53,500 °F)
b Top of inner layer of liquid hydrogen 11,000K (19,300 °F)
c Surface 120K (− 240 °F)

The rock temperature at the bottom of the deepest (12,600ft) mine on Earth, the Western Deep Levels Mine in S Africa's Transvaal, is 328K (131 °F). The hole temperature at the bottom of the 31,441ft-deep drilling in Washita County, Oklahoma, USA is 519K (475 °F).

EARTH 1

On these pages we compare temperatures at different heights within Earth's atmosphere and look at general patterns of world temperature distribution. Also included for comparison are some of the hottest and coldest temperatures ever recorded on Earth, ranging from 136.4°F (58°C) to −126.9°F (−88.3°C).

Lofty temperatures *right*
Indicated on this diagram are the temperature characteristics of regions within Earth's atmosphere. In the lowest region, the troposphere, temperature drops about 3°F (1.7°C) for every 1000ft (305m) of altitude. Temperature then rises in the stratosphere, falls in the mesosphere and rises again in the thermosphere. The wide range of temperature at the thermopause is accounted for by differences in solar activity, at night and during the day.

Curve and effect *left*
This diagram shows how temperature is affected by the curve of Earth's surface. In Polar regions the Sun's rays (**a,c**) are diffused over a greater area—and are therefore less effective—than they are near the Equator (**b**).

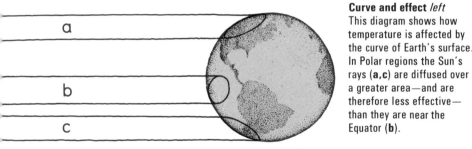

Ocean temperatures *left*
This diagram allows us to compare average annual mid-ocean surface temperatures at different latitudes. The curve of Earth's surface results in a general decrease in temperature toward the Poles (see diagram *above*), but other factors—such as the distribution of large land masses—modify the general pattern. Thus the highest temperatures are at 10°N and not at the Equator, and the decrease of temperature toward the Poles does not occur at a constant rate.

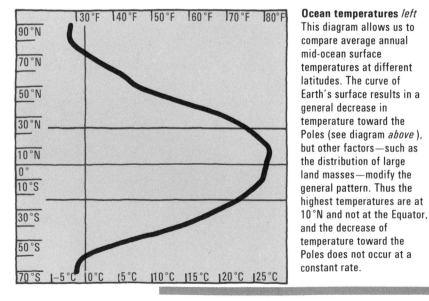

The East Sahara is the sunniest place on Earth, with an annual average of over 97% sunshine in daylight hours. The least sunny place is the North Pole, where no sun is recorded for winter stretches lasting 186 days.

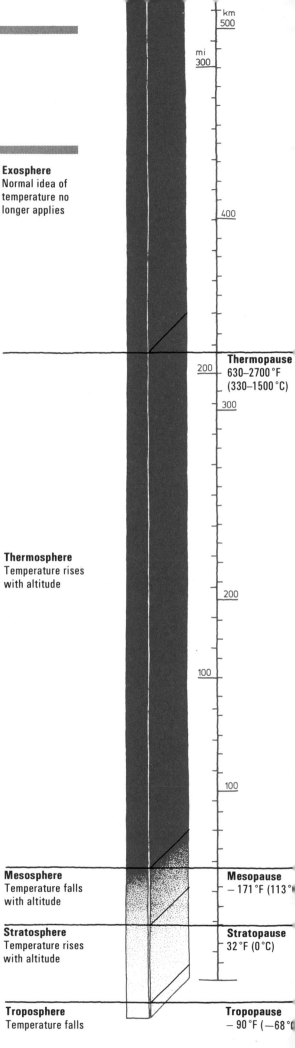

Exosphere
Normal idea of temperature no longer applies

Thermopause
630–2700°F
(330–1500°C)

Thermosphere
Temperature rises with altitude

Mesosphere
Temperature falls with altitude

Mesopause
−171°F (113°

Stratosphere
Temperature rises with altitude

Stratopause
32°F (0°C)

Troposphere
Temperature falls

Tropopause
−90°F (−68°C

The hottest place on Earth is Dallol in Ethiopia, with an annual average temperature of 94°F (34.4°C). The coldest place is Polus Nedostupnosti in Antarctica, with an annual average of −72°F (−57.8°C).

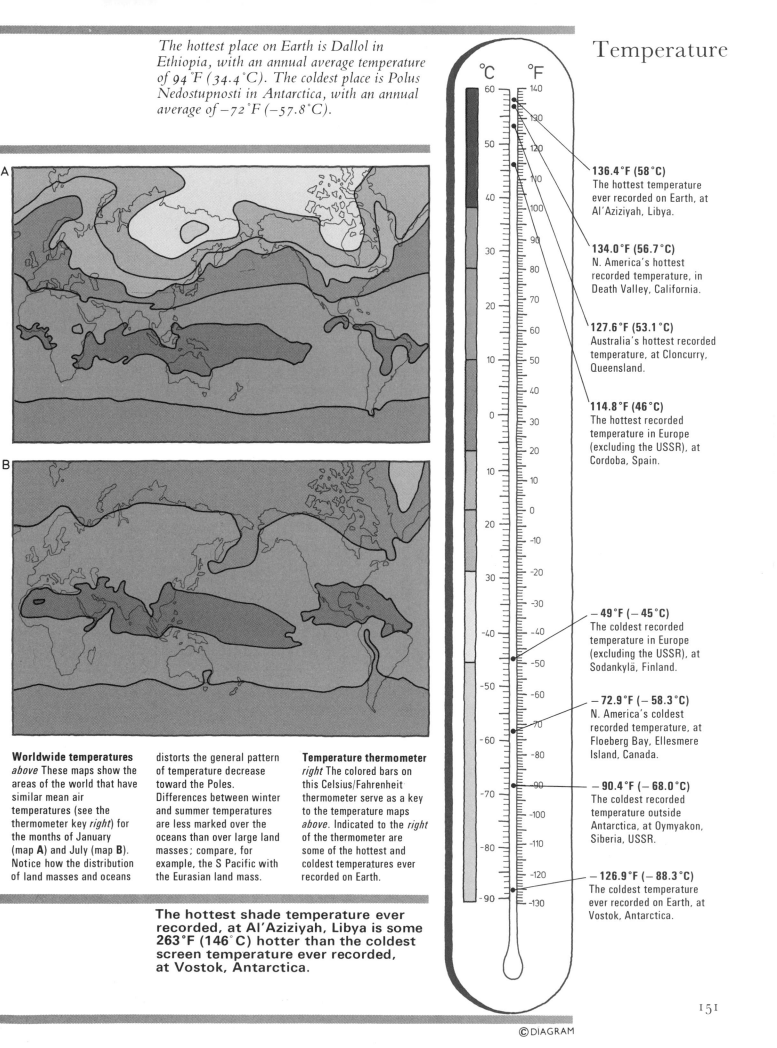

A

B

°C °F

136.4°F (58°C)
The hottest temperature ever recorded on Earth, at Al'Aziziyah, Libya.

134.0°F (56.7°C)
N. America's hottest recorded temperature, in Death Valley, California.

127.6°F (53.1°C)
Australia's hottest recorded temperature, at Cloncurry, Queensland.

114.8°F (46°C)
The hottest recorded temperature in Europe (excluding the USSR), at Cordoba, Spain.

−49°F (−45°C)
The coldest recorded temperature in Europe (excluding the USSR), at Sodankylä, Finland.

−72.9°F (−58.3°C)
N. America's coldest recorded temperature, at Floeberg Bay, Ellesmere Island, Canada.

−90.4°F (−68.0°C)
The coldest recorded temperature outside Antarctica, at Oymyakon, Siberia, USSR.

−126.9°F (−88.3°C)
The coldest temperature ever recorded on Earth, at Vostok, Antarctica.

Worldwide temperatures
above These maps show the areas of the world that have similar mean air temperatures (see the thermometer key *right*) for the months of January (map **A**) and July (map **B**). Notice how the distribution of land masses and oceans distorts the general pattern of temperature decrease toward the Poles. Differences between winter and summer temperatures are less marked over the oceans than over large land masses; compare, for example, the S Pacific with the Eurasian land mass.

Temperature thermometer
right The colored bars on this Celsius/Fahrenheit thermometer serve as a key to the temperature maps *above*. Indicated to the *right* of the thermometer are some of the hottest and coldest temperatures ever recorded on Earth.

The hottest shade temperature ever recorded, at Al'Aziziyah, Libya is some 263°F (146°C) hotter than the coldest screen temperature ever recorded, at Vostok, Antarctica.

EARTH 2

A city's annual temperature pattern depends first on latitude, and then on modifying factors such as its distance from the moderating influence of the sea and its altitude. Here we compare the monthly average temperatures and annual temperature ranges of representative cities from each continent.

City temperatures *below*
The graphs show monthly averages of maximum daily temperatures for selected cities in each continent. The cities are listed at the bottom of the page, where their latitudes are given to the nearest degree. Maps above the graphs show the cities' locations. Our selection illustrates the temperature patterns found in each continent. Mid- or high-latitude cities in the heart of large continents show the most pronounced temperature curves. Cities in low latitudes show the smallest monthly variations.

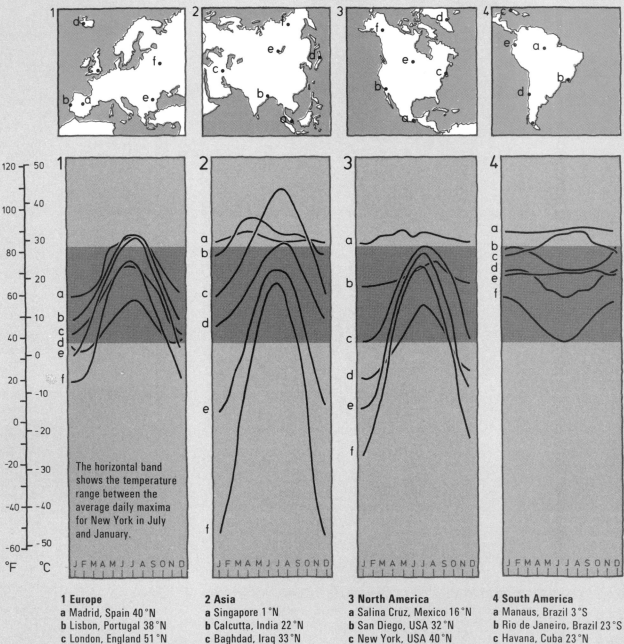

The horizontal band shows the temperature range between the average daily maxima for New York in July and January.

1 Europe
a Madrid, Spain 40 °N
b Lisbon, Portugal 38 °N
c London, England 51 °N
d Reykjavik, Iceland 64 °N
e Bucharest, Romania 44 °N
f Moscow, USSR 55 °N

2 Asia
a Singapore 1 °N
b Calcutta, India 22 °N
c Baghdad, Iraq 33 °N
d Tokyo, Japan 35 °N
e Irkutsk, USSR 52 °N
f Verkhoyansk, USSR 67 °N

3 North America
a Salina Cruz, Mexico 16 °N
b San Diego, USA 32 °N
c New York, USA 40 °N
d Angmagssalik, Greenland 65 °N
e Winnipeg, Canada 50 °N
f Dawson City, Canada 64 °N

4 South America
a Manaus, Brazil 3 °S
b Rio de Janeiro, Brazil 23 °S
c Havana, Cuba 23 °N
d Valparaiso, Chile 33 °S
e Quito, Ecuador 0 °
f Punta Arenas, Chile 53 °S

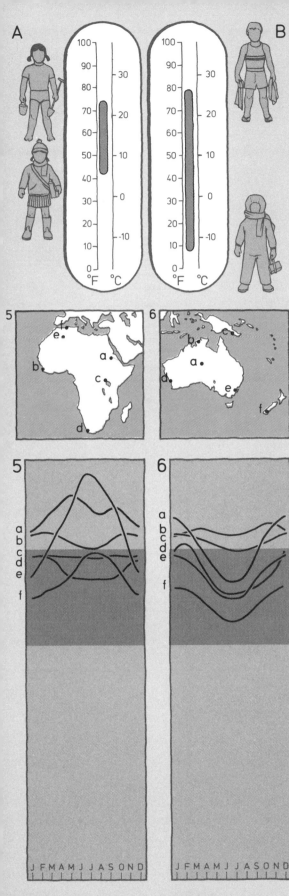

A

B

Temperature

Similar clothing can be worn on a typical summer's day in London (**A**) and in Winnipeg (**B**). But winter clothing suitable for London would be totally inadequate to cope with Winnipeg's average daily maximum temperatures in winter, even though the two cities are on similar latitudes.

City	Temperatures		Range	
2f Verkhoyansk	−54/65 °F	−47/18 °C	119 °F	65 °C
3f Dawson City	−16/72 °F	−27/22 °C	88 °F	49 °C
3e Winnipeg	7/78 °F	−14/26 °C	71 °F	40 °C
1f Moscow	20/75 °F	−7/24 °C	55 °F	31 °C
5e Ain Salah	69/117 °F	21/47 °C	48 °F	26 °C
3c New York	38/82 °F	3/28 °C	44 °F	25 °C
6a Alice Springs	67/98 °F	19/37 °C	31 °F	18 °C
1c London	42/73 °F	6/23 °C	31 °F	17 °C
1d Reykjavik	34/57 °F	1/14 °C	23 °F	13 °C
4f Punta Arenas	37/59 °F	3/15 °C	22 °F	12 °C
6e Sydney	59/79 °F	15/26 °C	20 °F	11 °C
2b Calcutta	78/96 °F	26/36 °C	18 °F	10 °C
5a Khartoum	89/107 °F	32/42 °C	18 °F	10 °C
6f Dunedin	48/65 °F	9/18 °C	17 °F	9 °C
5d Cape Town	68/80 °F	20/27 °C	12 °F	7 °C
4b Rio de Janeiro	72/82 °F	22/28 °C	10 °F	6 °C
5b Freetown	83/91 °F	28/33 °C	8 °F	5 °C
2a Singapore	85/89 °F	29/32 °C	4 °F	3 °C
4e Quito	68/71 °F	20/22 °C	3 °F	2 °C

Great and small ranges
The table *left* and the bar diagram *below left* show the annual temperature ranges of some of our cities, here ranked according to their size of range. The top of each bar marks the average daily maximum temperature for the hottest month; the bottom of each bar marks the same for the coldest month. Verkhoyansk has a much greater range (119 °F, 65 °C) than any other major weather station in any continent. Asia, S America and Africa all have stations with very small ranges (less than 5 °F/ 3 °C).

5

6

5

6

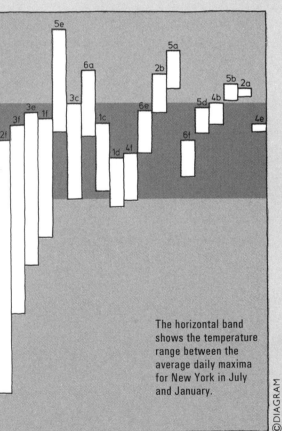

The horizontal band shows the temperature range between the average daily maxima for New York in July and January.

©DIAGRAM

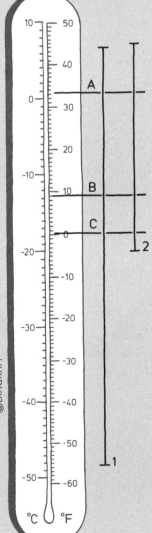

J F M A M J J A S O N D

J F M A M J J A S O N D

5 Africa
a Khartoum, Sudan 15 °N
b Freetown, Sierra Leone 8 °N
c Entebbe, Uganda 0°
d Cape Town, South Africa 33 °S
e Ain Salah, Algeria 27 °N
f Algiers, Algeria 36 °N

6 Oceania
a Alice Springs, Australia 23 °S
b Darwin, Australia 12 °S
c Port Moresby, Papua New Guinea 9 °S
d Perth, Australia 32 °S
e Sydney, Australia 34 °S
f Dunedin, New Zealand 46 °S

Extreme days *right*
Record movements of temperature are compared here with daily average temperatures for January.
1 At Browning, Montana, USA on January 23–24, 1916, the temperature fell 100 °F (55.5 °C), from 44 °F (6.7 °C) to −56 °F (−48.8 °C).

2 At Spearfish, S Dakota, USA, on January 22, 1943, the temperature rose 49 °F (27.2 °C) in 2 minutes, from −4 °F (−20 °C) to 45 °F (7.2 °C).
A New York 33.3 °F (0.7 °C).
B Moscow 9 °F (−12.7 °C).
C Winnipeg 0.1 °F (−17.7 °C)

153

BODY TEMPERATURES

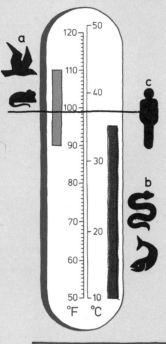

All animals need heat to keep their bodies alive. But some need or tolerate higher temperatures than others. Certain desert lizards can scamper over sun-baked rocks hot enough to fry an egg on. Built-in antifreeze enables the Antarctic ice fish to swim happily in water cold enough to turn ordinary blood to ice.

Temperature ranges *left*
The body temperatures of most warm-blooded animals fall within a fairly narrow range (**a**). The range of ideal temperatures for most cold-blooded animals is both wider and lower (**b**). The horizontal line marks man's normal temperature (**c**).

In a class of their own
Plotted *right* on the thermometers and listed in the tables are normal body temperatures of some warm-blooded animals (mammals **A**, birds **B**), and also ideal temperatures of some cold-blooded ones (reptiles **C**, amphibians and fish **D**).

A) Mammals

1	Goat	103.8 °F	39.9 °C
2	Domestic rabbit	101.3 °F	38.5 °C
3	Northern fur seal	99.9 °F	37.7 °C
4	Polar bear	99.1 °F	37.3 °C
5	Man	98.6 °F	37.0 °C
6	African elephant	97.5 °F	36.4 °C
7	Blue whale	95.9 °F	35.5 °C
8	Three-toed sloth	91.7 °F	33.2 °C
9	Spiny anteater	73.9 °F	23.3 °C

B) Birds

10	Western pewee	112.6 °F	44.8 °C
11	Canada jay	109.4 °F	43.0 °C
12	House sparrow	105.8 °F	41.0 °C
13	Wandering albatross	105.3 °F	40.7 °C
14	Owl	104.4 °F	40.2 °C
15	Hummingbird	104.2 °F	40.1 °C
16	Ostrich	102.6 °F	39.2 °C
17	King penguin	99.9 °F	37.7 °C
18	Arctic gull	93.2 °F	34.0 °C

A) Mammals *left*
Mammals have "thermostats" that help keep their bodies at a constant temperature. Heat produced by "burning" food is kept inside the body by fur. Surplus heat is lost by sweating, panting, or by convection, conduction, or radiation from the skin.

B) Birds *left*
Birds "burn" food fast to provide energy for flying, so their temperatures tend to be higher than mammals'. Of both sets of warm-blooded creatures listed, only one flying bird (**18**) is cooler than (**1**), the hottest-blooded mammal indicated.

Hibernation *left*
Shown are falls in body temperature survived by five warm-blooded creatures hibernating in cold weather. (Low body temperatures save energy and may prolong life when food is scarce.)
a Poor-will 104–64 °F
b Dormouse 98.6–35.6 °F
c Opossum 95–50.9 °F
d Common hamster 110–43 °F
e Marmot 107–50 °F
Hibernating mammals can raise their temperatures to avoid freezing. But hibernating amphibians seem immune to frostbite.

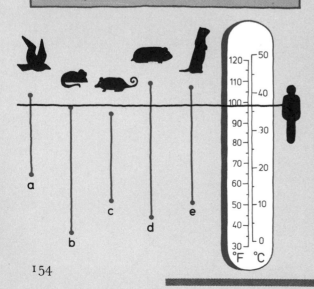

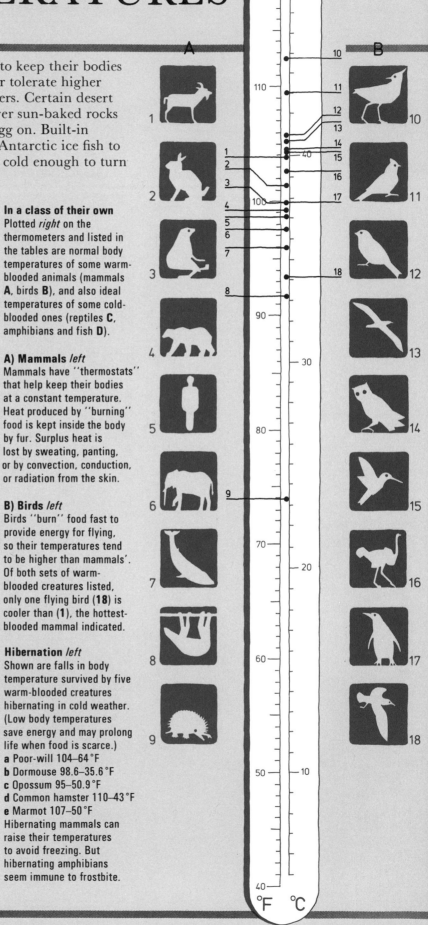

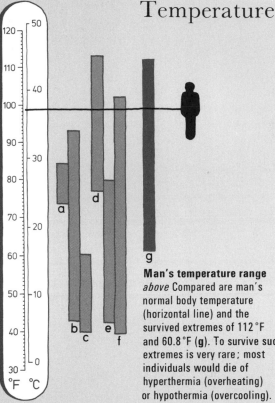

Temperature ranges *right*
Shown is the range of body temperature survived by six kinds of animal. Man's normal body temperature is included—as a horizontal line—for comparison.
a Crocodile 73.4–84.2 °F
b Catfish 42.8–93.2 °F
c Ascaphus frog 39.9–60.3 °F
d Horned lizard 77–113 °F
e Salamander 42.4–79.7 °F
f Garter snake 39.6–102 °F
Crocodiles tolerate only a small temperature change. Catfish and garter snakes survive the greatest fluctuations. The Ascaphus frog dies of overheating at temperatures that can kill a man with cold.

Man's temperature range
above Compared are man's normal body temperature (horizontal line) and the survived extremes of 112 °F and 60.8 °F (**g**). To survive such extremes is very rare; most individuals would die of hyperthermia (overheating) or hypothermia (overcooling).

C) Reptiles *right*
Many of these reptiles are most active when their body temperatures exceed 90 °F. But cold-blooded animals lack internal temperature controls. To keep an even temperature they must bask as the Sun rises, then hide in shade to cool down.

D) Amphibians/fish *right*
Most amphibians and fish are active at lower body temperatures than most reptiles. This applies especially to coldwater fish and to amphibians from cool climates. Most listed species thrive at below the given temperatures.

Environments *right*
a 104 °F (40 °C) desert habitat of rattlesnake.
b 93.9 °F (34.4 °C) Dallol, Ethiopia, the hottest town.
c 84.2 °F (29 °C) grassland habitat of red kangaroo.
d 69.8 °F (21 °C) lowest water temperature survived by coral-building polyps.
e 28.4 °F (−2 °C) average heat of Antarctic ocean surface, home of penguins.
f −58 °F (−50 °C) winter survived by arctic fox.
g −94 °F (−70 °C) winter survived by musk ox.
h −96 °F (−71.1 °C) record low at Oymyakon, USSR, coldest permanent town.

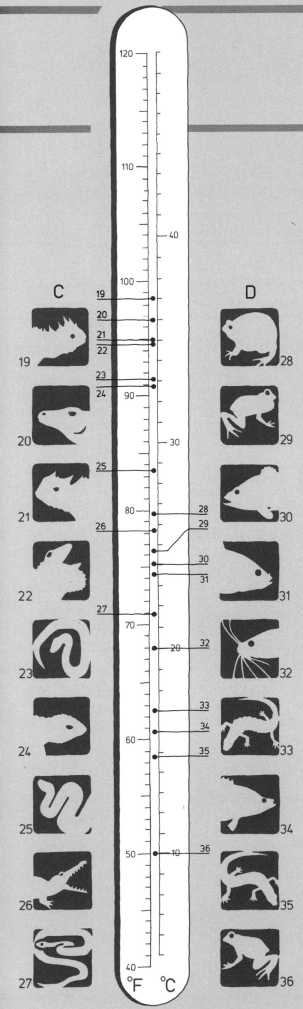

C) Reptiles		
19 Spiny lizard	98.4 °F	36.9 °C
20 Ornate lizard	96.8 °F	36.0 °C
21 Horned lizard	95.0 °F	35.0 °C
22 Australian bearded lizard	94.6 °F	34.8 °C
23 Nocturnal Saharan viper	91.4 °F	33.0 °C
24 Australian shingle-back skink	90.9 °F	32.7 °C
25 Indian python	83.5 °F	28.6 °C
26 Crocodile	78.1 °F	25.6 °C
27 Common garter snake	70.9 °F	21.6 °C

D) Amphibians/fish		
28 Mozambique rain frog	79.7 °F	26.5 °C
29 North American bullfrog	76.5 °F	24.7 °C
30 Large-mouthed black bass	75.2 °F	24.0 °C
31 Goldfish	74.3 °F	23.5 °C
32 Catfish	68.0 °F	20.0 °C
33 Mole salamander	62.6 °F	17.0 °C
34 Yellow perch	60.8 °F	16.0 °C
35 Lungless salamander	58.6 °F	14.8 °C
36 Ascaphus frog	50.0 °F	10.1 °C

©DIAGRAM

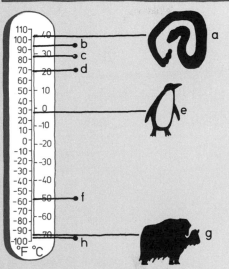

CHAPTER 8

DEGRES DES AC

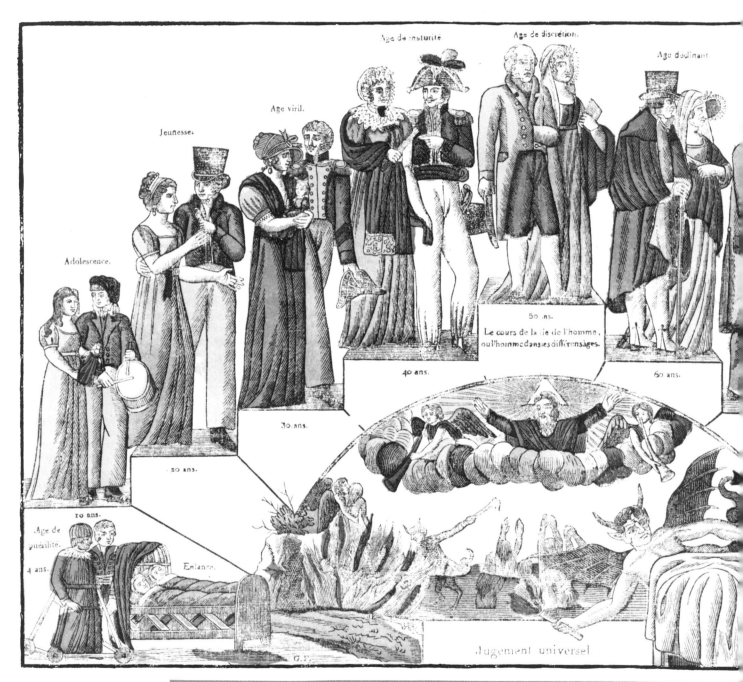

This 19th-century French popular print uses a staircase to demonstrate the stages in a couple's life. From infancy, they ascend through childhood, adolescence, youth, virility and maturity to reach a peak at the age of discretion. Thereafter they go into decline, to pass through decadence, frailty and decrepitude before returning again to infancy.

TIME

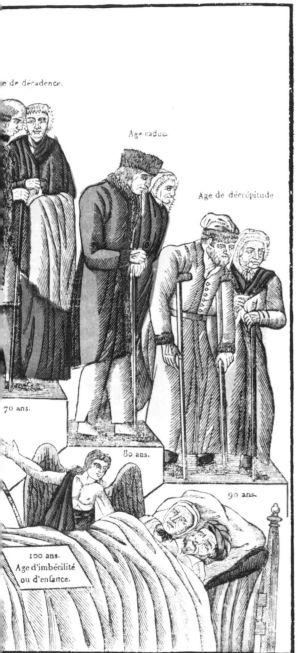

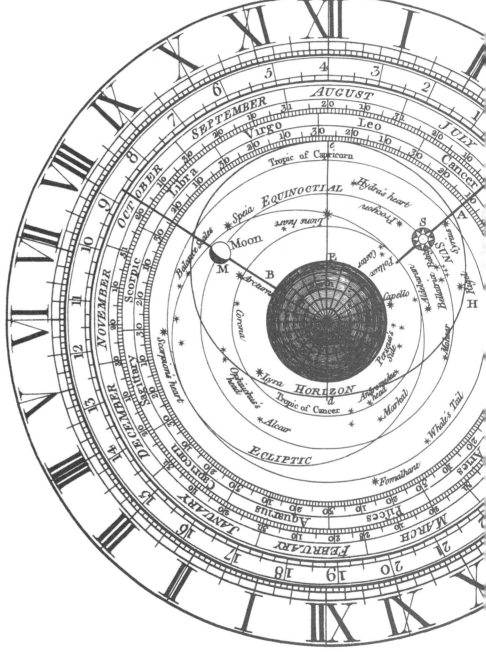

An illustration of an
elaborate clock face from
the second edition (1778)
of *Mechanical Exercises*,
an important work on
horology by the British
inventor James Ferguson.

MEASURING TIME 1

Time can be measured by motion, and it was the motion of Earth, Moon, Sun and stars that provided man with his first means of measuring time. Accurate lengths for the basic units of day, month and year were known thousands of years ago, even though the astronomical factors on which they were based were improperly understood.

Sidereal year	365.25636556 days	(365d 6h 9m 10s)
Anomalistic year	365.25964134 days	(365d 6h 13m 53s)
Tropical year	365.242198781 days	(365d 5h 48m 45s)
Sidereal month	27.32166 days	(27d 7h 43m 11s)
Tropical month	27.32158 days	(27d 7h 43m 5s)
Synodic month	29.53059 days	(29d 12h 44m 3s)
Mean solar day	1 day	(24h)
Sidereal day	0.997269 day	(23h 56m 4s)

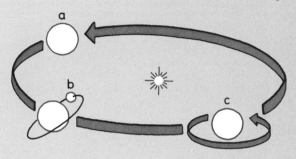

Years, months, days *left*
In simple terms, a year is the time it takes Earth to orbit once around the Sun (**a**), a month is the time it takes the Moon to make one orbit of Earth (**b**), and a day is the time it takes Earth to rotate once on its axis (**c**).

Complex timetable *above*
Different techniques of measurement give differing lengths for years, months and days. Sidereal times are calculated with reference to fixed stars. An anomalistic year requires Earth's orbit to be measured from the perihelion or aphelion (p. 29).

Tropical measurements refer to the apparent passage of the Sun and the actual passage of the Moon across Earth's equatorial plane. A synodic month is based on the phases of the Moon. A mean solar day relates to periods of darkness and light averaged over a year.

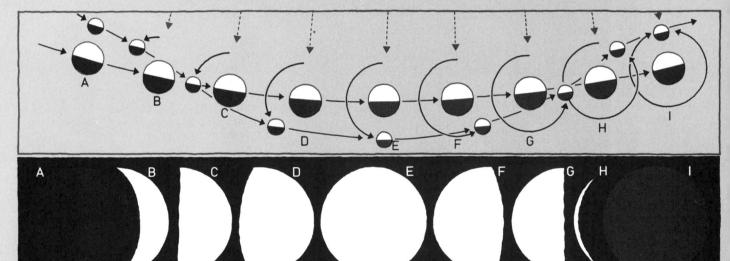

Synodic month *above*
Observation of the phases of the Moon has given man the time unit known as a synodic month. One full cycle of the Moon's phases takes approximately 29½ days. We illustrate how the Moon appears to us on selected nights within a month (**A–I**), and link this to a diagram showing the relative positions of Moon and Earth at different times in the month. The Moon always presents the same face to Earth, but the amount visible depends on how much of it is facing—and thus lit by—the Sun.

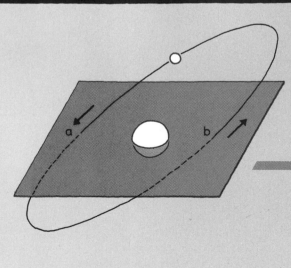

Tropical month *left*
The period of time known as a tropical month is approximately 27⅓ days long (compared to the 29½ days of a synodic month). As shown in the diagram, a tropical month is the period of time from when the orbiting Moon once passes through the plane of Earth's Equator to the next time that it passes through that plane in the same direction (**a** to **a** or **b** to **b**). It is measured with reference to the Moon's position in the sky, and is very close in length to a sidereal month.

There is a difference of more than two days between the length of a month measured by observing the phases of the Moon (a synodic month, 29½ days) and a month measured by noting the return of the Moon to a certain point in the sky (a tropical month, 27⅓ days).

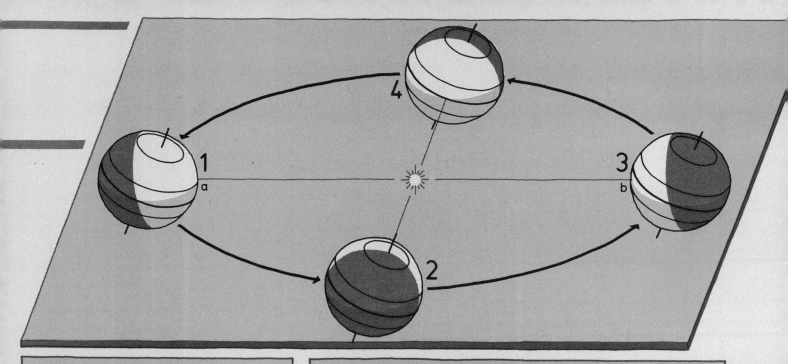

Northern hemisphere	Southern hemisphere
1 Summer	Winter
2 Autumn (fall)	Spring
3 Winter	Summer
4 Spring	Autumn (fall)

Date	Northern hemisphere	Southern hemisphere
1 June 21	Summer solstice	Winter solstice
2 September 23	Autumn equinox	Vernal (spring) equinox
3 December 22	Winter solstice	Summer solstice
4 March 21	Vernal (spring) equinox	Autumn equinox

©DIAGRAM

Years and seasons *above*
The period of approximately 365¼ days that we know as a year was discovered by careful observation of the Sun's apparent path through the sky. Our diagram shows Earth at four points in its orbit, corresponding to the seasons (see left-hand table).

Seasonal variations in insolation result from the inclination (23° 27') of Earth's axis to its plane of rotation around the Sun. Parts of the globe tilted away from the Sun receive less radiant energy per unit area than those receiving rays more directly (p. 150).

Equinox and solstice *above* The inclination of Earth to its plane of rotation around the Sun also produces variations in the relative lengths of ''day'' and ''night'' at different times of the year. The right-hand table beneath the diagram gives the dates

at which Earth is in the positions shown, and lists the solstices and equinoxes that occur on the different dates. Solstices are when the Sun appears to be overhead at midday at the maximum distance North and South of the Equator, currently on the two Tropics

(**a, b**). Days are longest and nights are shortest at the summer solstice, and vice versa at the winter solstice. At the equinoxes— when day and night are everywhere equal—the Sun appears directly overhead at midday at the Equator.

a 66° 33'N, 24 hours day
b 49° 3'N, 16 hours day
c Equator (0°), 12 hours day
d 49° 3'S, 8 hours day
e 66° 33'S, 0 hours day

Longest and shortest days
left Here we compare day lengths at different latitudes on June 21. On this date, places in the Northern hemisphere have their maximum, and places in the Southern hemisphere their minimum, number of hours of daylight.

Day by day *left*
A day measured by the Sun (ie the time between the Sun being at the same point overhead on two successive days, A_1–A_2) is longer than a sidereal day (the time between a distant star being at the same point on two successive days, A_1–B).

It could be said that the North and South Poles have only one ''day'' a year—made up of six months of daylight and six months of darkness.

MEASURING TIME 2

On the preceding two pages we looked at the astronomically based units of the day, the month and the year. Here we turn to "man-made" units, starting with multiples of years and days and going on to the various subdivisions of a day that have developed with the invention of increasingly accurate instruments for measuring time.

1 day (d)	= 24 hours (h, hr)
1 day	= 1440 minutes (m, min, ')
1 day	= 86,400 seconds (s, sec, '')
1 hour	= 0.0417 of a day
1 hour	= 60 minutes
1 hour	= 3600 seconds
1 minute	= 0.0006944 of a day
1 minute	= 60 seconds
1 second	= 0.0000115 of a day

Years and years *left*

Millennium	1000yr
Half-millennium	500yr
Century	100yr
Half-century	50yr
Decade	10yr
Half-decade	5yr

Listed here are some widely used names for periods of more than one year. Decade is derived from the Greek *deka* and the Latin *decem* meaning 10, century comes from the Latin *centuria*, whereas *millennium* is itself a Latin word for 1000 years.

Short and long weeks *right*
a 4-day intervals were formerly used in W Africa.
b The ancient Assyrians had weeks of 6 days.
c The 7-day week came to us from the ancient Babylonians and Jews.
d The French Revolutionary calendar had a 10-day week.

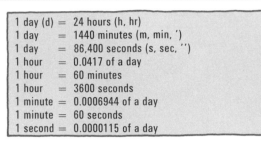

Hours in a day *left*
To measure the 24 hours in a day the hour hand of a conventional clock must make two full revolutions. Our diagram also shows the two systems of counting the hours, giving 24-hour clock equivalents for the morning and afternoon hours.

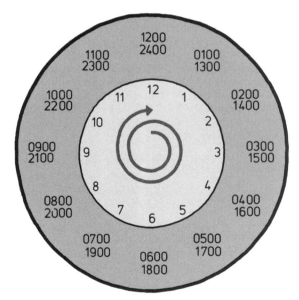

Minutes and seconds *right*
Here we use two dials to show the division of clock faces into the 60 sections used to measure minutes and seconds. The minute hand (**a**) makes one revolution in an hour, and the second hand (**b**) makes one in a minute and thus 60 in an hour.

Toward improved precision *right* Some important inventions in the history of time-keeping devices are listed here with their dates and then plotted against a logarithmic scale showing their average error in seconds or parts of a second per day.

a Mechanical clock c. 1280
b Pendulum clock c. 1650
c Mercury vial pendulum clock c. 1720
d Clocks with barometric compensation c. 1810
e Free-swinging pendulum clock c. 1910
f Quartz crystal clock 1929
g Cesium clock 1952

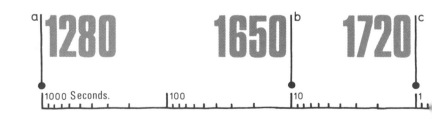

A medieval cook could get into a lot of trouble if he followed a modern recipe requiring him to cook a dish for an hour—for until the development of the mechanical clock, an "hour" was longer in summer than it was in winter. Whatever the relative length of daylight and darkness, the day was divided into 12 daytime and 12 nighttime hours.

A stopped clock shows the correct time on two occasions per day. This record is beaten only by a clock whose hands turn very much too quickly.

Units and equivalents *left*
This table gives the basic subdivisions of a day, with abbreviations in brackets the first time each unit appears. (The abbreviations h, m and s are officially preferred except where m might be confused for meters.)

Seconds of various lengths
If seconds are defined in terms of the length of a day (1/86400), then they are subject to the same variations as the day (see p. 158). Thus, for example, one sidereal second is approximately 0.997269 of a mean solar second.

A standard second *right*
The International System (SI) now takes the second as its base unit for time, defining it as 9,192,631,770 oscillations of a Cesium-133 atom. Here we give the equivalents of an SI second in mean solar seconds (**a**) and sidereal seconds (**b**).

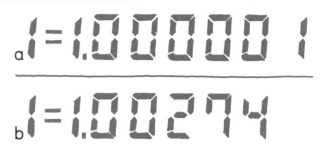

a $1 = 1.000001$

b $1 = 1.00274$

1 terasecond (Ts)	= 10^{12}s	= 31689 years
1 gigasecond (Gs)	= 10^9s	= 31.7 years
1 megasecond (Ms)	= 10^6s	= 11.6 days
1 kilosecond (ks)	= 10^3s	= 16.67 minutes
1 millisecond (ms)	= 10^{-3}s	= 0.001s
1 microsecond (μs)	= 10^{-6}s	= 0.000001s
1 nanosecond (ns)	= 10^{-9}s	= 0.000000001s
1 picosecond (ps)	= 10^{-12}s	= 0.000000000001s
1 femtosecond (fs)	= 10^{-15}s	= 0.000000000000001s
1 attosecond (as)	= 10^{-18}s	= 0.000000000000000001s

Table of seconds *left*
As this table indicates, a consequence of greater precision in the measuring of seconds has been the application to them of offical metric suffixes. Names are here followed by their abbreviations and equivalents.

A decimal clock *right*
Here we show an unusual clock face (**1**) devised to measure time according to a decimalized time system. Although unlikely to catch on, it does provide some interesting comparisons. The clock has three hands.
A The deciday hand makes one revolution each day, marking off 10 decidays (each equal to 2.4 hours).
B The centiday hand makes 10 revolutions in a day, each of them marking off 10 centidays (each 14.4 minutes) and 100 millidays.
C The fastest hand makes 1000 revolutions a day, one each milliday (86.4 seconds).

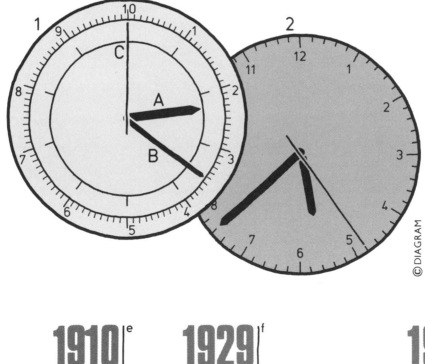

© DIAGRAM

Two-timing *left*
The hands on these two clock faces show the same time of day. The decimal clock (**1**) expresses it as 2 decidays 3 centidays and 5 millidays (see explanation to *left* of the clock faces). The hands on the conventional clock (**2**) show the equivalent time of 05h 38m 24s.

1810 d

1910 e

1929 f

1952 g

|0·1 |0·01 |0·001 |0·0001 |0·00001 |0·000001

A clock that is losing 30 minutes a day will show the correct time once every 24 days, but a clock that is losing only one-thousandth of a second per day will be correct only once every 118,275 years.

CALENDARS

Most calendars represent an attempt to measure easily the days and parts of days counted in observations of the changing positions in the sky of the Sun or Moon (see pages 158–159). Here we make comparisons between different calendars and also look at some of the events used as starting points when counting years.

A	Jewish religion's date for the Creation	3761BC
B	Starting point for the Mayan ''Long Count''	3111BC
C	Era named for Kali, consort of the god Siva	3102BC
D	Introduction of current Chinese year system	c.1600BC
E	Date for the founding of the city of Rome	753BC
F	Date taken to be that of the Buddha's birth	544BC
G	Date adopted as that of the birth of Christ	1AD
H	Traditional date for the flight of Mohammed	622AD
I	Declaration of the 1st French Republic	1792AD

The beginnings of time
The table *above* and the diagram *below* show years— reckoned in Christian or Common Era terms—that have been taken as starting points for the counting of years. Most are dates of religious significance, but others (**D, E, I**) have secular origins. Some are still in use (**A, D, F, G, H**), but others lost favor (**C, E, I**), or died with a civilization (**B**).

A

C

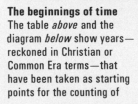

B

D

Year upon year *left*
Comparisons for 1980–81AD.
a 1980AD, starts January 1, lasts 366 days.
b Chinese year, starts our February 16, lasts 355 days.
c Jewish year, starts our September 11, lasts 383 days.
d Muslim year, starts our November 9, lasts 355 days.

Year of Our Lord 1980
right English names for the months are followed here by the number of days that they contain in 1980, a 366-day ''leap'' year. The calendar system described here is now more widely used than any other. Known as the Gregorian calendar, it is a 16th-century adaptation of the Julian calendar devised in the 1st century BC. By it:
a) years whose number is not divisible by 4 have 365 days;
b) centennial years, eg 1900, have 365 days unless the figures before the noughts are exactly divisible by 4;
c) other years have 366 days.

✝ 1980

January	(31)
February	(29)†
March	(31)
April	(30)
May	(31)
June	(30)
July	(31)
August	(31)
September	(30)
October	(31)
November	(30)
December	(31)

† 28 days except in leap years

Jewish and Muslim years
right Here we list in order the names of the months (with numbers of days in brackets) for the Jewish year 5741 and the Muslim year 1401, both of which start during 1980AD. A Jewish year has 13 not 12 months if its number, when divided by 19, leaves 0, 3, 6, 8, 11, 14 or 17. Its precise number of days is fixed with reference to particular festivals that must not fall on certain days of the week. A Muslim year has 355 not 354 days if its number, when divided by 30, leaves 2, 5, 7, 10, 13, 16, 18, 21, 24, 26 or 29.

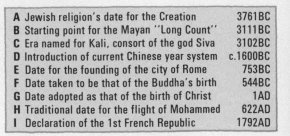

The Gregorian calendar, devised in the 16th century, is considerably more accurate than its predecessor, the Julian calendar. Over a 4000-year period, the Gregorian calendar loses only 1 day 4 hours and 55 minutes when compared to the tropical year. The Julian calendar loses 31 days 4 hours and 13 minutes over a similar period.

Years short and long *right*
Here we show how the number of days in a year varies among cultures and from year to year.
a Years with 365 or 366 days derive from the solar year (see p. 158), which lasts roughly 365¼ days.
b The Muslim year, based on 12 lunar cycles each of approximately 29½ days, has a total of 354 or 355 days.
c The Jewish year is also lunar, but to keep broadly in line with the solar cycle some years have 12 months (353, 354 or 355 days) and others have 13 months (383, 384 or 385 days).

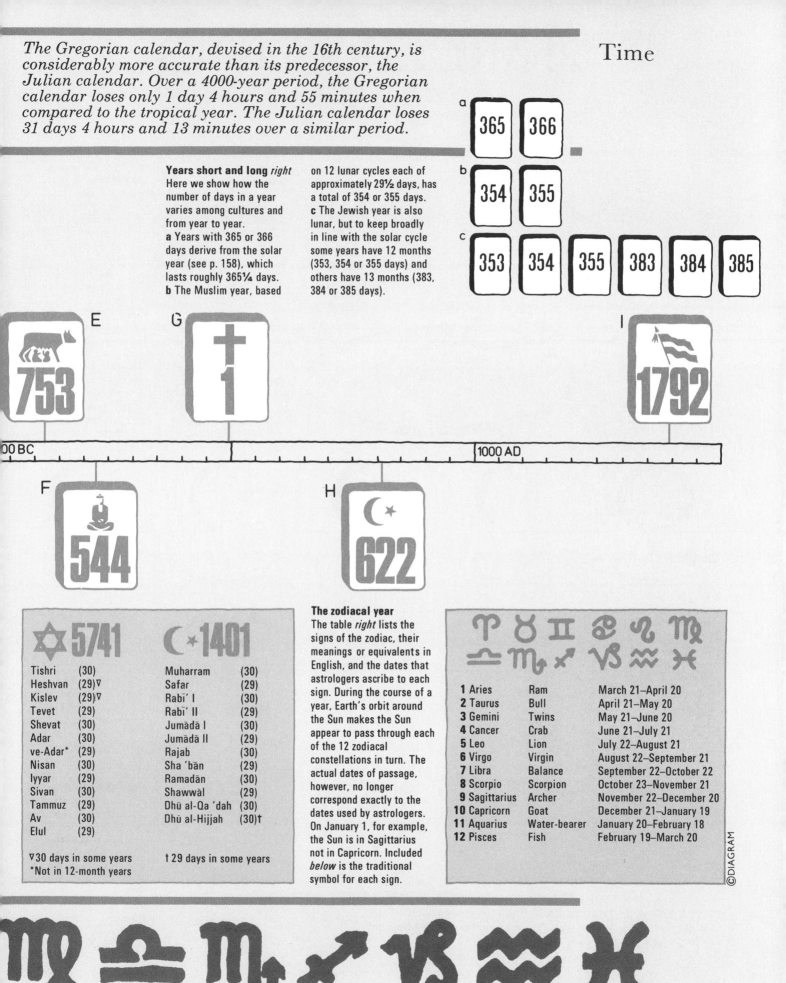

The zodiacal year
The table *right* lists the signs of the zodiac, their meanings or equivalents in English, and the dates that astrologers ascribe to each sign. During the course of a year, Earth's orbit around the Sun makes the Sun appear to pass through each of the 12 zodiacal constellations in turn. The actual dates of passage, however, no longer correspond exactly to the dates used by astrologers. On January 1, for example, the Sun is in Sagittarius not in Capricorn. Included *below* is the traditional symbol for each sign.

1	Aries	Ram	March 21–April 20
2	Taurus	Bull	April 21–May 20
3	Gemini	Twins	May 21–June 20
4	Cancer	Crab	June 21–July 21
5	Leo	Lion	July 22–August 21
6	Virgo	Virgin	August 22–September 21
7	Libra	Balance	September 22–October 22
8	Scorpio	Scorpion	October 23–November 21
9	Sagittarius	Archer	November 22–December 20
10	Capricorn	Goat	December 21–January 19
11	Aquarius	Water-bearer	January 20–February 18
12	Pisces	Fish	February 19–March 20

5741

Tishri	(30)	Muharram	(30)
Heshvan	(29)▽	Safar	(29)
Kislev	(29)▽	Rabī' I	(30)
Tevet	(29)	Rabī' II	(29)
Shevat	(30)	Jumādā I	(30)
Adar	(30)	Jumādā II	(29)
ve-Adar*	(29)	Rajab	(30)
Nisan	(30)	Sha 'bān	(29)
Iyyar	(29)	Ramadān	(30)
Sivan	(30)	Shawwāl	(29)
Tammuz	(29)	Dhū al-Qa 'dah	(30)
Av	(30)	Dhū al-Hijjah	(30)†
Elul	(29)		

1401

▽30 days in some years   †29 days in some years
*Not in 12-month years

PLANETARY TIMES

We on Earth are very accustomed to our day of approximately 24 hours and our year of approximately 365 days. A look at the days and years of other planets, however, shows that they keep very different time. Venus, for example, has a day longer than its year, whereas a single year on Neptune consists of over 90,000 of its days.

Planet	Rotation period	Sidereal period
A Mercury	59 days	88 days
B Venus	243 days (retrograde)	224.7 days
C Earth	23h 56m 4s	365.256 days
D Mars	24h 37m 23s	687 days
E Jupiter	9h 50m 30s	11.86 years
F Saturn	10h 14m	29.46 years
G Uranus	11h (retrograde)	84.01 years
H Neptune	16h	164.8 years
I Pluto	6 days 9h	247.7 years

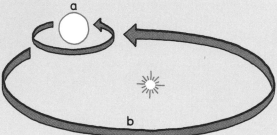

Days and years *left*
The diagram shows a planet's day (**a**) and year (**b**). A planet's day—properly called its rotation period—is the time taken by the planet to rotate once on its axis. Its year—or sidereal period—is the time it takes to orbit the Sun once.

All the planets *above*
Listed are the rotation period (day) and sidereal period (year) for each of the planets, in distance order from the Sun. Times are in sidereal Earth days and years (see p. 158). Retrograde means rotating E–W relative to fixed stars.

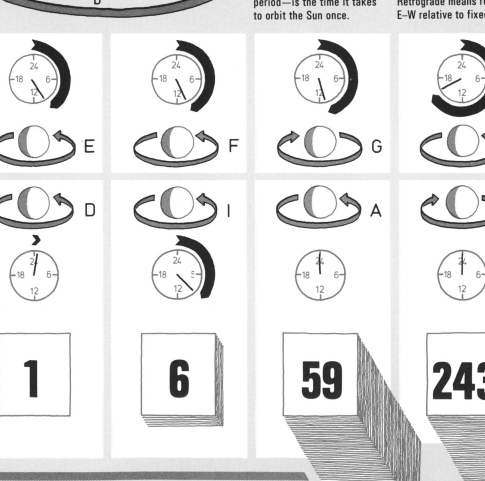

Shorter days *above, right*
Here we use 24h-clock faces to compare the length of day on Earth (**C**) with the lengths of days on planets with shorter days:
Jupiter (**E**) 9h 50m 30s
Saturn (**F**) 10h 14m
Uranus (**G**) 11h
Neptune (**H**) 16h

Longer days *right*
Clock faces and calendar pages show the lengths of days on planets with days that are longer than an Earth day:
Mars (**D**) 24h 37m 23s
Pluto (**I**) 6 days 9h
Mercury (**A**) 59 days
Venus (**B**) 243 days

Earth (a) rotates on its axis 25.38 times for every once that the Sun (b) rotates on its own axis. Thus it could be argued that only one in 25.38 Earth days is truly a "Sun-day."

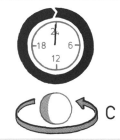

**A child on Pluto would have to wait
247 Earth years for his first birthday.
A child on Mercury, however,
would have a much better deal—
four birthdays in a single Earth year!**

Years and years and years
right This diagram provides
a visual comparison of year
lengths on the planets (as
listed *left*). Thick lines on
the diagram indicate how
many of their own years—
or what portion of them—
the various planets experience
for every Earth year (**C**).

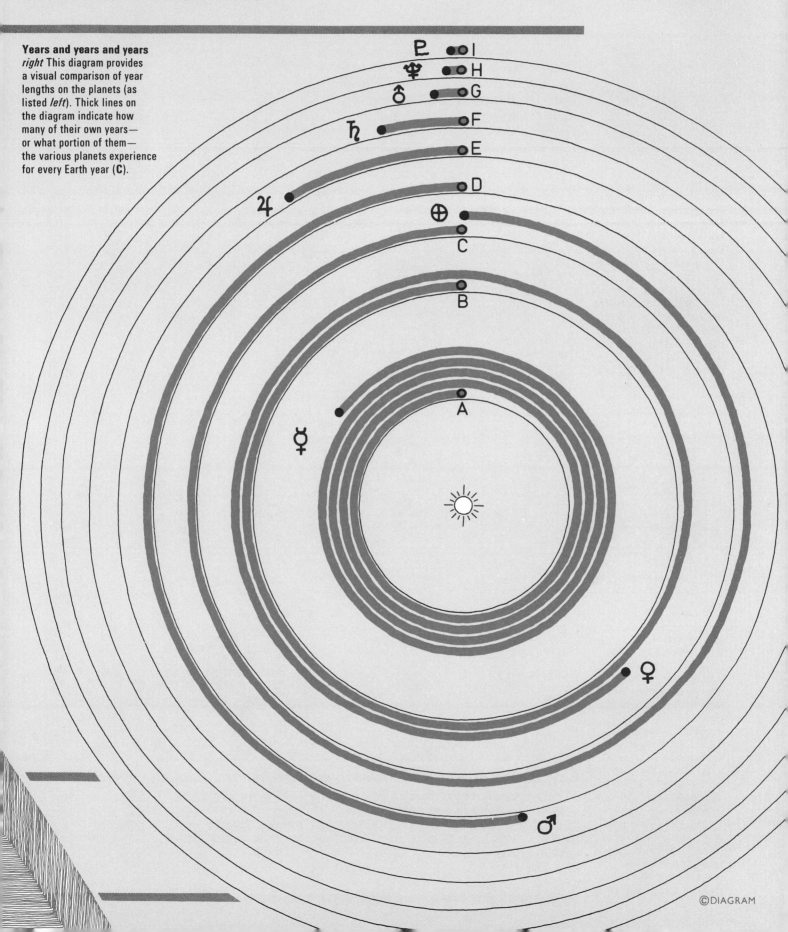

©DIAGRAM

TIME ZONES

True noon at any place is when the Sun is at its highest point in the sky. If time were reckoned solely by the Sun, it would be necessary to turn the clock forward by 4 minutes for every degree of latitude traveled east. In practice, however, it has been more convenient to develop a system of internationally recognized time zones.

More time for summer
In order to benefit from more daylight in summer some countries in higher latitudes adopt Daylight Saving Time. Clocks are advanced, usually 1 hour, in spring and put back in the fall. (Our map ignores these adjustments.)

World time zones *right*
The map shows standard times, relative to Greenwich Mean Time (GMT). Each country chooses the standard time, or times, most convenient to it, so modifying the basic pattern whereby the globe is divided into time zones each 15° of longitude wide. In principle, successive zones to the east of the Greenwich zone (centered on the Greenwich Meridian, marked **A** on the map) are 1 hour in advance of GMT, and successive zones west of it are 1 hour behind GMT. The International Date Line is marked **B** on the map.

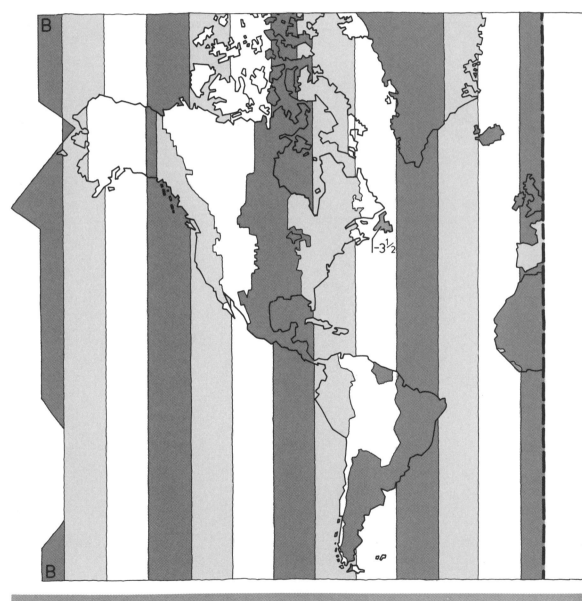

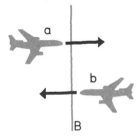

Making a date *above*
If two aircraft, one of them flying east (**a**) and the other west (**b**), were to set off from the Greenwich Meridian (0° longitude) and then fly on to see each other at the other side of the world (180° longitude), passengers in aircraft **a** would have gained 12 hours whereas those in aircraft **b** would have lost 12 hours. In theory there would be a whole day between them. In practice, however, the problem is solved by the International Date Line, where the date officially changes (**B** on the map *right*).

When New Yorkers are eating breakfast (8 a.m. local time), Londoners are eating lunch (1 p.m.), Muscovites are taking tea (4 p.m.), Djakartans are eating dinner (8 p.m.), while in Wellington it is time for a bedtime drink (midnight).

In the USSR all trains and airplanes keep Moscow time (GMT +3). Thus the station clock at Nakhodka, the Eastern terminus of the Trans-Siberian Railway, shows a time seven hours in advance of local time (GMT +10).

London to L.A. in 1 hour !
right The regular flying time between London and Los Angeles is 9 hours, but if we consider only local times passengers arrive 1 hour after take-off. The explanation lies in the 8-hour difference between the cities' standard times.

1 London local time:
a Take-off 1 p.m.
b Arrival 10 p.m.
2 Los Angeles local time:
a Take-off 5 a.m.
b Arrival 2 p.m.

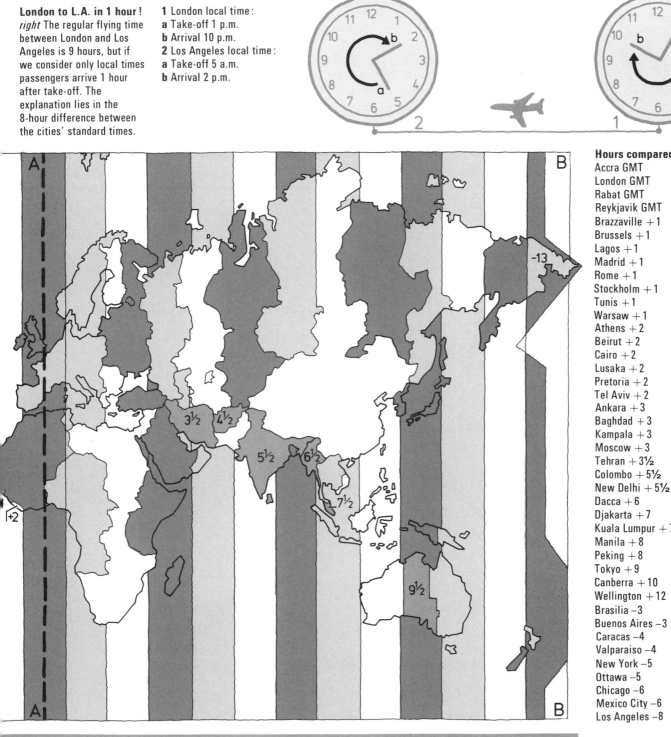

Hours compared to GMT

City	GMT offset
Accra	GMT
London	GMT
Rabat	GMT
Reykjavik	GMT
Brazzaville	+1
Brussels	+1
Lagos	+1
Madrid	+1
Rome	+1
Stockholm	+1
Tunis	+1
Warsaw	+1
Athens	+2
Beirut	+2
Cairo	+2
Lusaka	+2
Pretoria	+2
Tel Aviv	+2
Ankara	+3
Baghdad	+3
Kampala	+3
Moscow	+3
Tehran	+3½
Colombo	+5½
New Delhi	+5½
Dacca	+6
Djakarta	+7
Kuala Lumpur	+7½
Manila	+8
Peking	+8
Tokyo	+9
Canberra	+10
Wellington	+12
Brasilia	−3
Buenos Aires	−3
Caracas	−4
Valparaiso	−4
New York	−5
Ottawa	−5
Chicago	−6
Mexico City	−6
Los Angeles	−8

Concorde flies from London to New York in an amazing three hours, allowing a London businessman to arrive in New York some two hours earlier than the time he set off!

GEOLOGICAL TIME

The relative ages of Earth's rocks can be determined from fossils within them and from the relative positions of distinctive rock layers. But only with the development of techniques based on the rate of decay of radioactive substances did it become possible to give reasonably accurate dates for the various geological time divisions.

Geological eras *right*
This scale diagram shows the comparative durations and approximate dates of the four geological eras.
A Precambrian (4600 million to 570 million years ago). It contains two periods:
1 Archean (4600 million to 2600 million years ago);

2 Proterozoic (2600 million to 570 million years ago).
B Paleozoic (570 million to 230 million years ago).
C Mesozoic (230 million to 65 million years ago).
D Cenozoic (65 million years ago to the present).

Era	Period	Epoch
A Precambrian	**1** Archean	
	2 Proterozoic	
B Paleozoic	**3** Cambrian	
	4 Ordovician	
	5 Silurian	
	6 Devonian	
	7 Carboniferous*	
	8 Permian	
C Mesozoic	**9** Triassic	
	10 Jurassic	
	11 Cretaceous	
D Cenozoic	**12** Paleogene	**a** Paleocene
		b Eocene
		c Oligocene
	13 Neogene	**d** Miocene
		e Pliocene
	14 Quaternary	**f** Pleistocene
		g Holocene

*Equivalents in N America are:
7i Mississippian
7ii Pennsylvanian

Geological time divisions
The table *above* shows the division of geological time into eras, periods and epochs. The spiral diagram *top right* shows the initial division into eras (**A, B, C, D**). The three diagrams *right*, drawn to different scales, illustrate the approximate duration and dates of subdivisions within the three most recent eras. Period divisions (**1–14**) are keyed on the side of the bars; epoch divisions (**a–g**) are keyed on the top. The pictures above the bars are of animals that lived when those rocks were laid down.

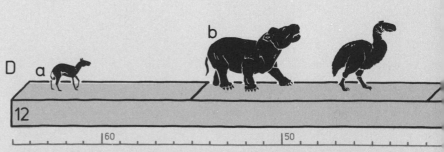

"The moving creature that hath life" was, according to the account in Genesis, created on the fifth day of the Creation. Scientists consider that the first primitive organisms appeared some 3500 million years ago, 1100 million years after the formation of our planet.

The Precambrian era spans some 4030 million years, seven times longer than the time spanned by the other three geological eras together.

A1

A2

B   C   D

900 800 700 600 500 400 300 200 100

5   6   7   8

7i   7ii

400 350 300 250

11

160 150 140 130 120 110 100 90 80 70

d   e   f   g

13   14

30   20   10   0

The present geological epoch, the Holocene, began approximately 10,000 years ago. It therefore accounts for a mere 0.00021% of our geological time scale.

EVOLUTION

Here we show some of the creatures that lived in the three most recent eras of geological time. The boxes contain a selection of animals that illustrate the evolution of animal life from primitive arthropods to the appearance of early man. As on the previous two pages all dates given are for millions of years ago.

Evolution in a year *right*
The diagram shows Earth's evolution in terms of one 365-day year. Man's time on Earth is seen to be relatively short—ape-man appeared at 18.17 on December 31 and Christ was born only 14 seconds before the year end.

a Jan 1, Earth formed
b Mar 29, life begins
c Nov 16, trilobites common
d Nov 27, first fish
e Dec 4, first amphibians
f Dec 15, first dinosaurs
g Dec 26, dinosaurs die out
h Dec 28, mammals diversify
i Dec 31, first ape-man
j Dec 31, birth of Christ

Creatures of the past
Illustrated *right* and listed *below* is a representative selection of animals from the three most recent eras of geological time.

A) Paleozoic era (570–230)
Traces of life are present in older Precambrian rocks, but an abundance of fossils is first found in the Paleozoic era. Trilobites were especially numerous throughout this era, dying out only in the Permian period (280).

B) Mesozoic era (230–65)
The first mammal and the first bird appeared during the Mesozoic era. There was a great increase in the number of reptiles and land animals. Dinosaurs were dominant until the end of the Cretaceous period (65).

C) Cenozoic era (65–0)
Throughout the Cenozoic era mammals diversified and developed rapidly. Animal life in general began to assume the forms we recognize today. Primitive man first appeared only in the last three million years of this era.

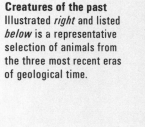

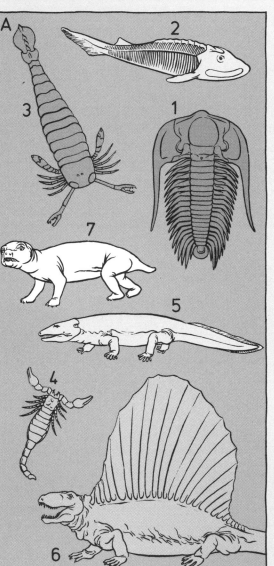

A) Paleozoic era
1 Trilobite, arthropod, common in Cambrian period 570–500
2 Ostracoderm, jawless fish of Ordovician period 500–436
3 Eurypterid, sea scorpion of Silurian period 436–396
4 Scorpion, early land animal of Silurian period 436–396
5 Icthyostega, early amphibian of Devonian period 396–346
6 Dimetrodon, reptile of early Permian period 280–225
7 Thrinaxodon, Permian mammal-like reptile 280–225

B) Mesozoic era
a Pantothere, the first mammal, Triassic period 230–195
b Ammonite, cephalopod, common from Triassic to late Cretaceous period 230–65
c Archaeopteryx, the first bird, Jurassic period 195–141
d Stegosaurus, armored, herbivorous dinosaur, Jurassic period 195–141

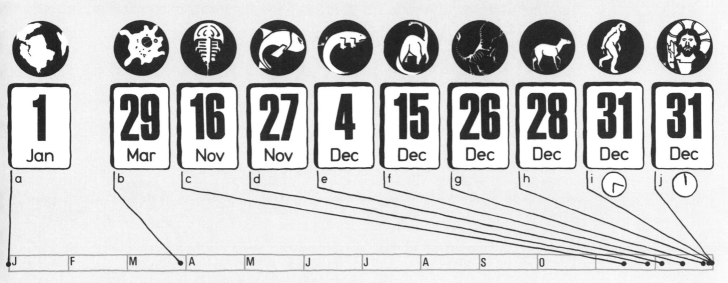

| 1 Jan (a) | 29 Mar (b) | 16 Nov (c) | 27 Nov (d) | 4 Dec (e) | 15 Dec (f) | 26 Dec (g) | 28 Dec (h) | 31 Dec (i) | 31 Dec (j) |

J F M A M J J A S O

C

©DIAGRAM

e Plesiosaur, marine reptile, common from Jurassic to Cretaceous period 195–65

f Pteranodon, flying reptile of Cretaceous period 141–65

g Tyrannosaurus, carnivorous dinosaur of Cretaceous period 141–65

h Triceratops, heavily armored, three–horned dinosaur of late Cretaceous period 100–65

C) Cenozoic era

1 Eohippus, ancestor of horse, Eocene epoch 55–42

2 Diatryma, flightless bird of Eocene epoch 55–42

3 Mesohippus, ancestor of horse, Oligocene epoch 42–22.5

4 Paleomastodon, ancestor of elephant, Oligocene epoch 42–22.5

5 Australopithecus, ape-man of Pliocene epoch, c.3.0

6 Smilodon, saber-toothed tiger, Pleistocene epoch 1.8–0.01

7 Woolly mammoth of Pleistocene epoch 1.8–0.01

GESTATION AND INCUBATION

From the fertilization of an egg cell to birth or hatching takes hours or many months, according to the species involved. An elephant's possible 730 days' gestation is 61 times longer than the shortest known opossum pregnancy. A fertilized new-laid tortoise egg needs up to 114 days to hatch, a fruit fly's less than 1 day.

Mammals *right*
Shown are average pregnancy times for 10 mammals. Most small kinds, and all whose young are born tiny and undeveloped, have short pregnancies. Pregnancy is longest in big mammals like the rhinoceros, which bears large well-formed young.

1 Common opossum 13 days
2 House mouse 19 days
3 Dog 63 days
4 Goat 151 days
5 Chimpanzee 237 days
6 Man 265 days
7 Camel 406 days
8 Giraffe 410 days
9 Rhinoceros 560 days
10 Indian elephant 624 days

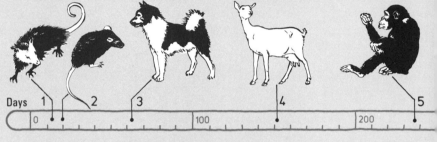

Birds *right*
This diagram shows average incubation periods for 10 birds. Eggs are fertilized and start developing in the body, but once laid they stop developing until incubation begins. Some birds sit on their eggs seven times as long as others.

A Finch 12 days
B Thrush 14 days
C Wren 16 days
D Pheasant 21.5 days
E Falcon 28.5 days
F Swan 30 days
G Ostrich 42 days
H Hawk 44 days
I Emperor penguin 63 days
J Royal albatross 79 days

Reptiles *right*
Shown are incubation and gestation periods for 10 reptiles. Some skinks and snakes bear living young*, but most reptiles lay eggs incubated by the Sun—the period depending on heat received and development, if any, before eggs are laid.

1 Australian skink 30 days
2 Grass lizard 42 days
3 Marine turtle 55 days
4 Hog-nosed snake 60 days
5 Alligator 61 days
6 Python 61.5 days
7 Spiny lizard 63 days
8 Box turtle 87 days
9 Viviparous lizard 90 days*
10 Tortoise 105 days

Insect metamorphosis
right This diagram shows average numbers of days spent by four insects in different developmental stages from new-laid egg to becoming adults. Most have three stages (egg, larva, pupa), but some have only two (egg, larva).

A Honeybee (worker): egg 3, larva 7, pupa 11 days
B Fruit fly: egg less than 1, larva 7, pupa 5 days
C Silkmoth: egg 10, larva 23, pupa 17 days
D Louse: egg 13, larva 8 days, no pupal stage

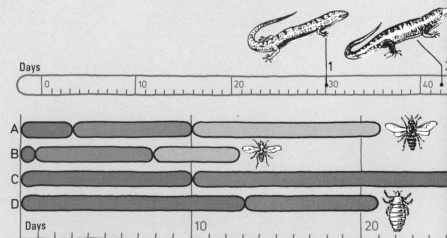

A rattlesnake father would have quite a problem if he wanted to be present at the hatching of his offspring—for incubation takes anything from 140 to 295 days.

The typical gestation period for an Indian elephant is 624 days —33 times as long as that for a house mouse, and 48 times as long as that for a common opossum.

Fast developers *left*
Given here are examples of very brief incubation periods or pregnancy terms among insects, birds, mammals and reptiles.
a Fruit fly less than 1 day
b Some finches 10 days
c Common opossum 13 days
d Some skinks 30 days

Slow developers *right*
This diagram shows some of the longest-known periods of incubation or pregnancy among birds, insects, reptiles and mammals.
a Royal albatross 81 days
b Some butterflies 270 days
c Tuatara 425 days
d Indian elephant 730 days

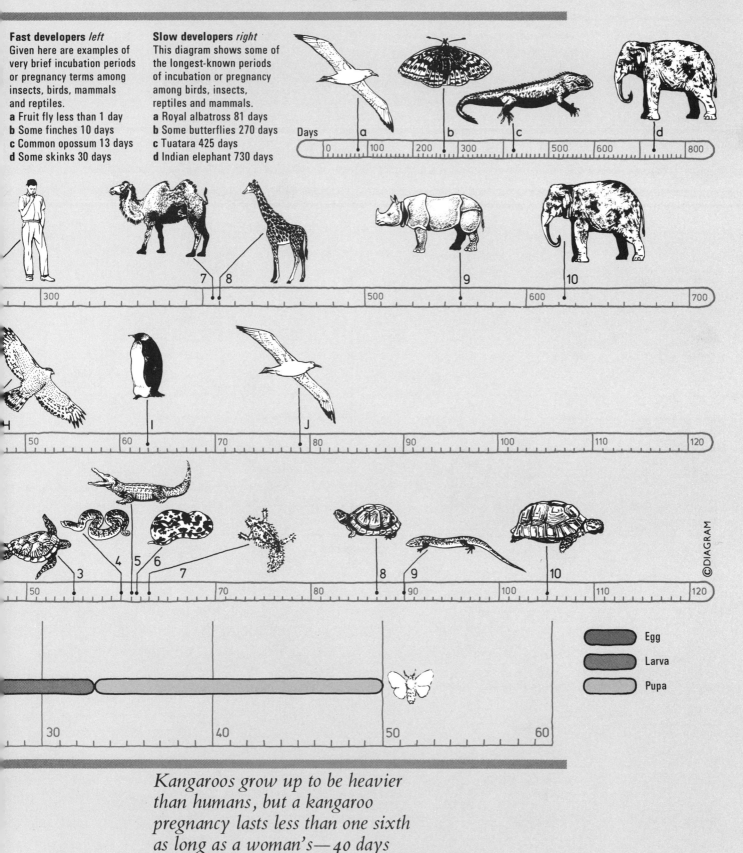

Days
a b c d
0 100 200 300 400 500 600 700 800

7 8 9 10
300 500 600 700

H I J
50 60 70 80 90 100 110 120

3 4 5 6 7 8 9 10
50 70 80 90 100 110 120

©DIAGRAM

Egg
Larva
Pupa

30 40 50 60

Kangaroos grow up to be heavier than humans, but a kangaroo pregnancy lasts less than one sixth as long as a woman's—40 days compared with 265 days.

LIFE EXPECTANCIES

Man is often acutely aware of the span of his own existence, and yet he has one of the longest life spans of all animals. Medical advances now allow more people to live to an old age, so increasing average life expectancies at birth. Women have a greater natural life expectancy than men, but to offset this more males than females are born.

Animals and man *below*
Only bacteria live longer than the tortoise, for which a life span of 100 years is nothing out of the ordinary. Man's average life expectancy of around 70 years in developed countries exceeds even record figures for most other animals.

Some typical life spans are indicated on this diagram and listed *right*.

a Mayfly (imago) 1 day
b Mouse 2–3yr
c Trout 5–10yr
d Squirrel 11yr
e Rabbit 12yr
f Sheep 10–15yr
g Cat 13–17yr
h Rattlesnake 18yr
i Owl 24yr
j Lion 25yr
k Horse 30yr
l Hippopotamus 40yr
m Pelican 45yr
n Ostrich 50yr
o Alligator 55yr
p African elephant 60yr
q Macaw 63yr
r Dolphin 65yr
s Raven 69yr
t Rhinoceros 70yr
u Man 68yr, woman 76yr (USA)
v Tortoise 100yr

© DIAGRAM

If the fox hadn't got him first, the old gray goose might have lived to be 31 years old.

"And all the days of Methuselah were nine hundred sixty and nine years: and he died." (Genesis V 27)

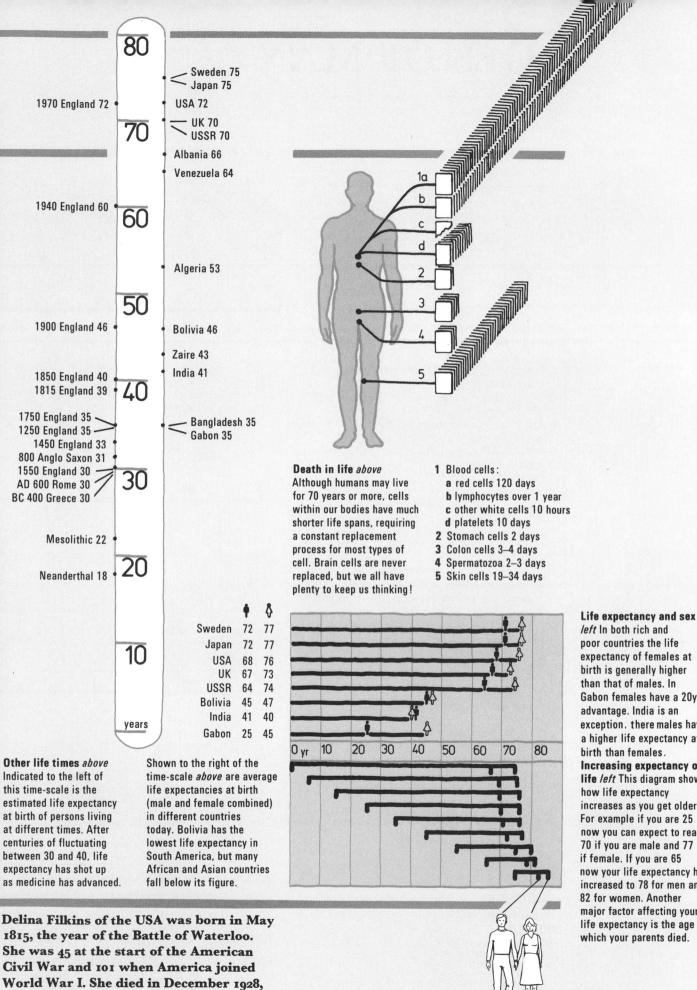

Timeline scale (years, left axis):

- Sweden 75
- Japan 75
- 1970 England 72 · USA 72
- UK 70 / USSR 70
- Albania 66
- Venezuela 64
- 1940 England 60
- Algeria 53
- 1900 England 46 · Bolivia 46
- Zaire 43
- 1850 England 40 · India 41
- 1815 England 39
- 1750 England 35 · Bangladesh 35
- 1250 England 35 · Gabon 35
- 1450 England 33
- 800 Anglo Saxon 31
- 1550 England 30
- AD 600 Rome 30
- BC 400 Greece 30
- Mesolithic 22
- Neanderthal 18

Death in life *above*
Although humans may live for 70 years or more, cells within our bodies have much shorter life spans, requiring a constant replacement process for most types of cell. Brain cells are never replaced, but we all have plenty to keep us thinking!

1 Blood cells:
  a red cells 120 days
  b lymphocytes over 1 year
  c other white cells 10 hours
  d platelets 10 days
2 Stomach cells 2 days
3 Colon cells 3–4 days
4 Spermatozoa 2–3 days
5 Skin cells 19–34 days

Life expectancy and sex *left* In both rich and poor countries the life expectancy of females at birth is generally higher than that of males. In Gabon females have a 20yr advantage. India is an exception, there males have a higher life expectancy at birth than females.

Increasing expectancy of life *left* This diagram shows how life expectancy increases as you get older. For example if you are 25 now you can expect to reach 70 if you are male and 77 if female. If you are 65 now your life expectancy has increased to 78 for men and 82 for women. Another major factor affecting your life expectancy is the age at which your parents died.

Life expectancy by sex and country:

	♂	♀
Sweden	72	77
Japan	72	77
USA	68	76
UK	67	73
USSR	64	74
Bolivia	45	47
India	41	40
Gabon	25	45

0 yr 10 20 30 40 50 60 70 80

Other life times *above*
Indicated to the left of this time-scale is the estimated life expectancy at birth of persons living at different times. After centuries of fluctuating between 30 and 40, life expectancy has shot up as medicine has advanced.

Shown to the right of the time-scale *above* are average life expectancies at birth (male and female combined) in different countries today. Bolivia has the lowest life expectancy in South America, but many African and Asian countries fall below its figure.

Delina Filkins of the USA was born in May 1815, the year of the Battle of Waterloo. She was 45 at the start of the American Civil War and 101 when America joined World War I. She died in December 1928, aged 113 years 214 days.

AGES OF MAN

Growth, maturation and aging take up various proportions of life in different people. But a life span of 70 years from the time of conception might consist roughly of the following: 1% pre-birth development and growth; 3% infancy; 14% childhood; 9% adolescence; 31% prime of life; 29% middle age; 13% old age.

A (Day 1) Sperm fertilizes ovum released into mother's oviduct.

B (Day 7) Fertilized ovum has formed a blastocyst embedded in the uterus.

C (Day 13) The blastocyst has produced a yolk sac and embryonic disk.

D (Day 23) Embryo now has the makings of heart, brain and spinal cord.

E (Day 29) Embryo now plainly has a head, and buds that will be arms and legs.

F (Week 5) Embryo about 0.5in (1.3cm); eyes, ears, vital organs taking shape.

G (Week 9) Embryo about 2in (5cm) long and weighs about 0.35oz (10gm).

H (Week 14) Embryonic development has given way to fetal growth. The fetus is 7in (18cm) and weighs 4oz (113gm).

I (Week 26) At 15in (38cm) and 2lb (907gm) fetus may survive if born.

J (Week 38) Baby born, measuring 20in (51cm) and weighing 7lb (3.18kg).

K (1 year old) Infant says first words, crawls, and can almost stand and walk.

L (3 years) Child knows own age in years, walks erect, easily climbs stairs.

M (5 years) Runs on toes, does useful chores, begins to read and write.

N (7 years) Learns to ride a bicycle, tell the time, draw a man side-face.

O (8 years) Solves simple mathematical problems; enjoys reading.

P (10 years) Solves problems in less concrete situations than before.

Q (13 years) Puberty has arrived, also early adolescent growth spurt.

R (18 years) Adolescence ending and early adulthood beginning.

S (40 years) The slow decline of faculties called middle age sets in.

T (60 years) Increasing adverse bodily changes mark the onset of old age.

U (75 years) Advanced old age sees over 50% loss of certain basic faculties.

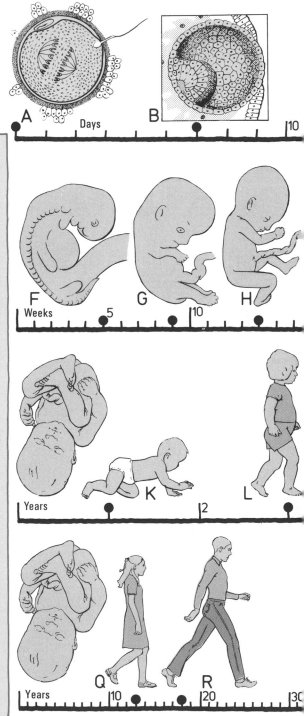

The days of our years are threescore years and ten (Psalm 90).

And one man in his time plays many parts, His acts being seven ages (William Shakespeare, "As You Like It").

Development accounts for only the first 14 weeks (36%) of a typical 38-week pregnancy.

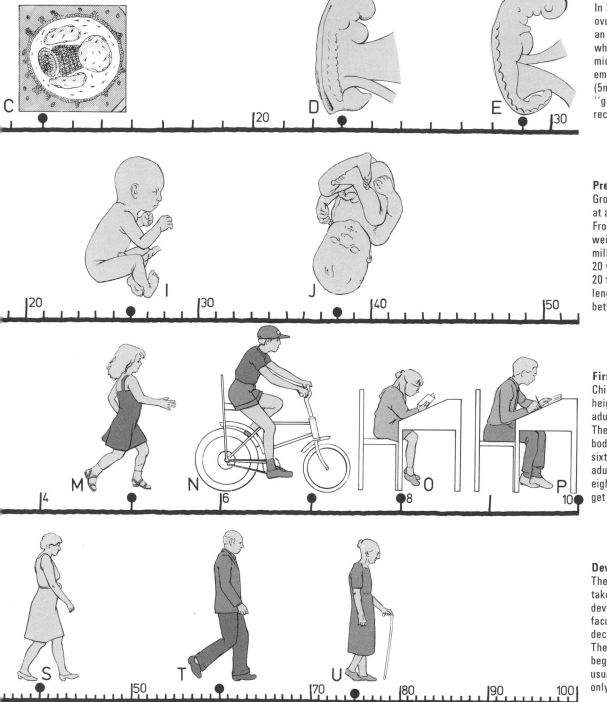

First 30 days
In 30 days the fertilized ovum divides repeatedly at an amazing rate. By day 30 what had begun as a microscopic speck is an embryo more than 0.2in (5mm) long, but tail and ''gill clefts'' confusingly recall an embryo amphibian.

Pre-birth growth
Growth is faster now than at any later time in life. From fertilization to birth, weight increases 5000 million times; in the next 20 years it goes up only 20 times more. Crown-heel length increases 12½ times between 8 weeks and birth.

First 10 years
Children reach half adult height by 2 years and half adult weight by about 10. The head is one-quarter body length at birth, one-sixth by 6 (one-eighth when adult); legs start three-eighths body length but get proportionately longer.

Development and decline
The first 18 years or so are taken up by growth and development. But some faculties are already in decline by adolescence. The main aging process has begun by the late 20s but usually becomes apparent only considerably later.

If age 60 is taken as the starting point for old age, then Japan's Shigechiyo Izumi—114 last birthday—has been ''old'' for 48% of his life. A typical person's life span of 70 years includes only 13% old age.

FAMOUS LIVES

The length of a person's life does not necessarily affect the measure of his achievement. Some famous persons have had comparatively short lives, others have lived for a long time but been active or in positions of power for only a small portion of their total life span.

A	John Keats, English romantic poet (1795–1821)	25
B	Wilfred Owen, English war poet (1893–1918)	25
C	Georges Seurat, French artist (1859–91)	31
D	Franz Schubert, Austrian composer (1797–1828)	31
E	W.A. Mozart, Austrian composer (1756–91)	35

Short and long lives
Plotted on the time scale *right* are the ages at which some famous writers, artists and composers died. Those with short lives are listed in the table *above* the scale, and those with long lives in the table *below* it.

Lives of Presidents *right*
Here we give dates of life and term of office, age on taking office (**a**) and on death (**b**) of US Presidents: longest lived (**1**), oldest into office (**2**), first (**3**), youngest to die of natural causes (**4**), and youngest elected and to die (**5**).

Lives of monarchs
Shown in the diagram *right* are the life spans and reigns of some short- and long-lived monarchs. The tables *below* the diagram list each monarch's dates of life and reign, length of reign (**a**), and age at death (**b**). John I of France, the posthumous son of Louis X, was born a King but survived only a few days. Edward V, one of the Princes in the Tower, was King of England for only 77 days. Excepting unreliably documented claims, Louis XIV had the longest reign of any monarch.

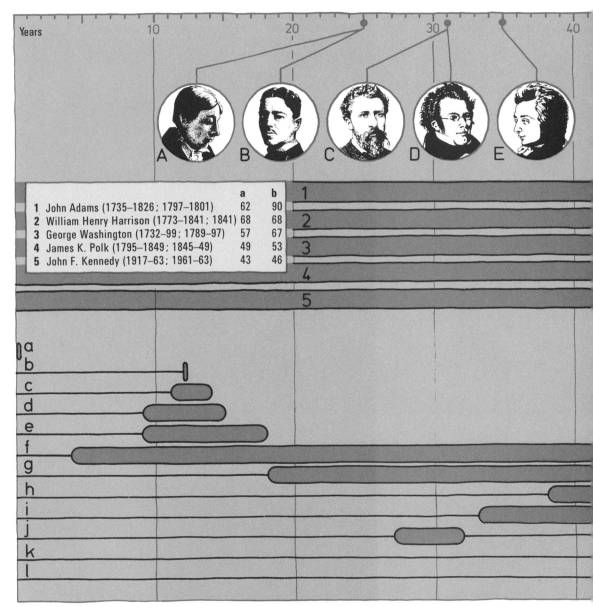

		a	b
1	John Adams (1735–1826; 1797–1801)	62	90
2	William Henry Harrison (1773–1841; 1841)	68	68
3	George Washington (1732–99; 1789–97)	57	67
4	James K. Polk (1795–1849; 1845–49)	49	53
5	John F. Kennedy (1917–63; 1961–63)	43	46

		a	b
a	John I of France (1316; 1316)	0.01	0.01
b	Edward V of England (1470–83; 1483)	0.21	12
c	Peter II of Russia (1715–30; 1727–30)	3	14
d	Edward VI of England (1537–53; 1547–53)	6	15
e	Tutankhamun of Egypt (c.1370–52BC; c.1361–52BC)	9	18

Louis XIV was King of France for 95% of his lifetime, reigning for 72 out of his 76 years. Stanislaus I of Poland lived to be 12 years older than Louis XIV, but of his total of 88 years he was King for only 5 years, a mere 6% of his lifetime.

Short but famous lives
left Listed here, with the dates between which they lived and the ages at which they died, are people who, despite the shortness of their lives, made major contributions as writers, artists or composers.

Long and memorable lives
below Included in this table, with their dates and the ages at which they died, is a selection of famous writers, artists and composers who are also noteworthy for their exceptionally long lives.

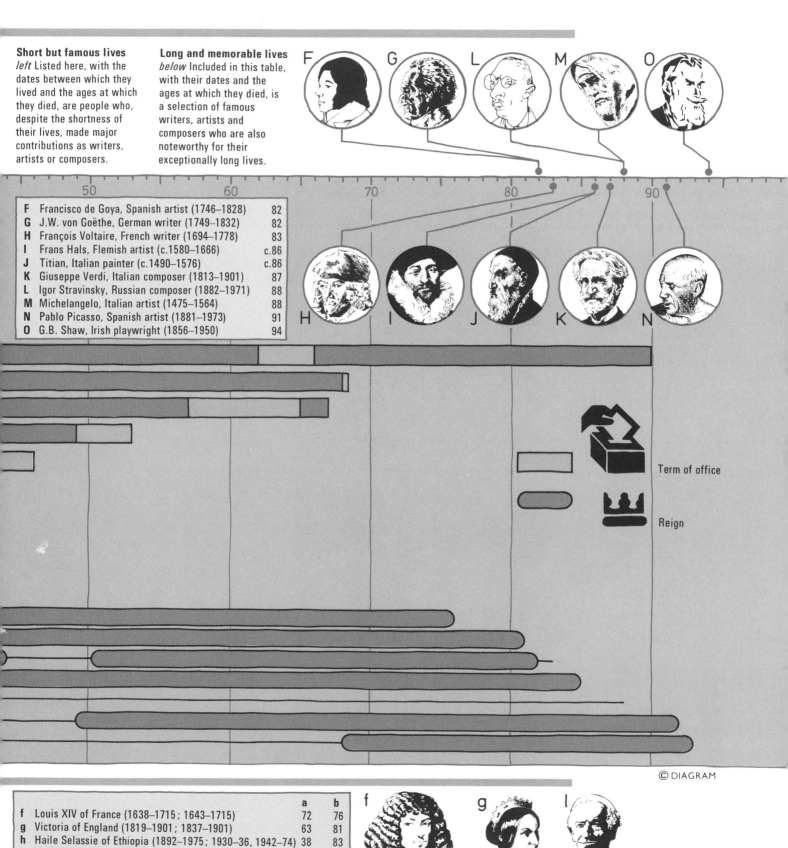

F	Francisco de Goya, Spanish artist (1746–1828)	82
G	J.W. von Goëthe, German writer (1749–1832)	82
H	François Voltaire, French writer (1694–1778)	83
I	Frans Hals, Flemish artist (c.1580–1666)	c.86
J	Titian, Italian painter (c.1490–1576)	c.86
K	Giuseppe Verdi, Italian composer (1813–1901)	87
L	Igor Stravinsky, Russian composer (1882–1971)	88
M	Michelangelo, Italian artist (1475–1564)	88
N	Pablo Picasso, Spanish artist (1881–1973)	91
O	G.B. Shaw, Irish playwright (1856–1950)	94

Term of office

Reign

© DIAGRAM

		a	b
f	Louis XIV of France (1638–1715; 1643–1715)	72	76
g	Victoria of England (1819–1901; 1837–1901)	63	81
h	Haile Selassie of Ethiopia (1892–1975; 1930–36, 1942–74)	38	83
i	Haakon VII of Norway (1872–1957; 1905–57)	51	85
j	Stanislaus I of Poland (1677–1766; 1704–09)	5	88
k	Gustav V of Sweden (1858–1950; 1907–50)	42	92
l	Pope Leo XIII (1810–1903; 1878–1903)	25	93

179

EVENTS IN HISTORY

From the start of the Bronze Age in the Middle East through to the present day is a period of approximately 5000 years. Here we use a visual presentation of selected dates within this period to show the relationships in time between one important event and another and between these same events and the present day.

a	Bronze Age in Syria and Palestine	c.3000BC	4980
b	Sumerian Classical Age begins	c.2850BC	4830
c	Egyptian Old Kingdom begins	c.2615BC	4595
d	Harappa culture in Indus Valley	c.2000BC	3980
e	Height of Minoan civilization	c.1600BC	3580
f	Shang dynasty in China	c.1400BC	3380
g	Israelites ruled by King David	1000BC	2980
h	Golden Age of Athens	c.450BC	2430
i	Rome defeats Carthage	146BC	2126
*	Birth of Christ		c.1980

Ancient civilizations
Indicated on the time scale *left* and listed in the table *above* are key dates from civilizations that pre-dated the birth of Christ. The final column in the table shows the number of years between each event and 1980.

Traveling back in time
The time spiral *right* takes us back in time from the year 2000AD to the birth of Christ. Major events and famous reigns are indicated on the spiral, and listed *below* together with their dates and number of years back from 1980.

A	First men on the Moon	1969	11
B	World War 2	1939–45	35
C	World War 1	1914–18	62
D	American Civil War	1861–65	115
E	Latin American revolutions	1806–25	155
F	Napoleon, ruler of France	1799–1815	165
G	French Revolution	1789	191
H	American independence	1776	204
I	Frederick the Great of Prussia	1740–86	194
J	Louis XIV of France	1643–1715	265
K	Thirty Years' War	1618–48	332
L	Pilgrims arrive at Cape Cod	1620	360
M	Queen Elizabeth I of England	1558–1603	377
N	Reformation started by Luther	1517	463
O	Columbus crosses the Atlantic	1492	488
P	Lorenzo de' Medici (of Florence)	1478–92	488
Q	Black Death in Europe	1348–50	630
R	Genghis Khan	1206–27	753
S	Third Crusade	1189–92	788
T	Norman Conquest of England	1066	914
U	Important Viking expeditions	840–885	1095
V	Emperor Charlemagne	771–814	1166
W	Traditional start of Islam	622	1358
X	Pope Gregory the Great	590–604	1376
Y	End of Roman Empire in West	476	1504
Z	Roman Empire at maximum extent	117	1863
*	Birth of Christ		c.1980

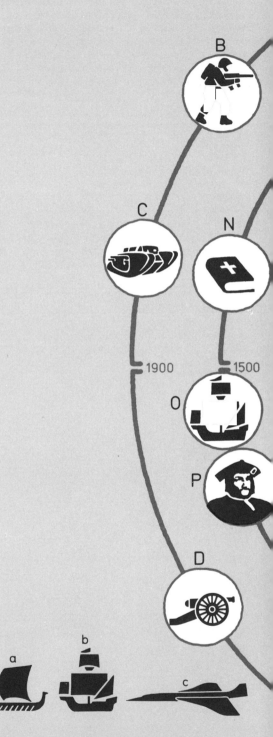

Leif Ericsson (a) reputedly crossed the Atlantic in c. 1000AD, roughly 500 years before Columbus's first crossing (b) in 1492. A similar time span separates Columbus's voyage from the first transatlantic passenger flight of Concorde (c) in 1977.

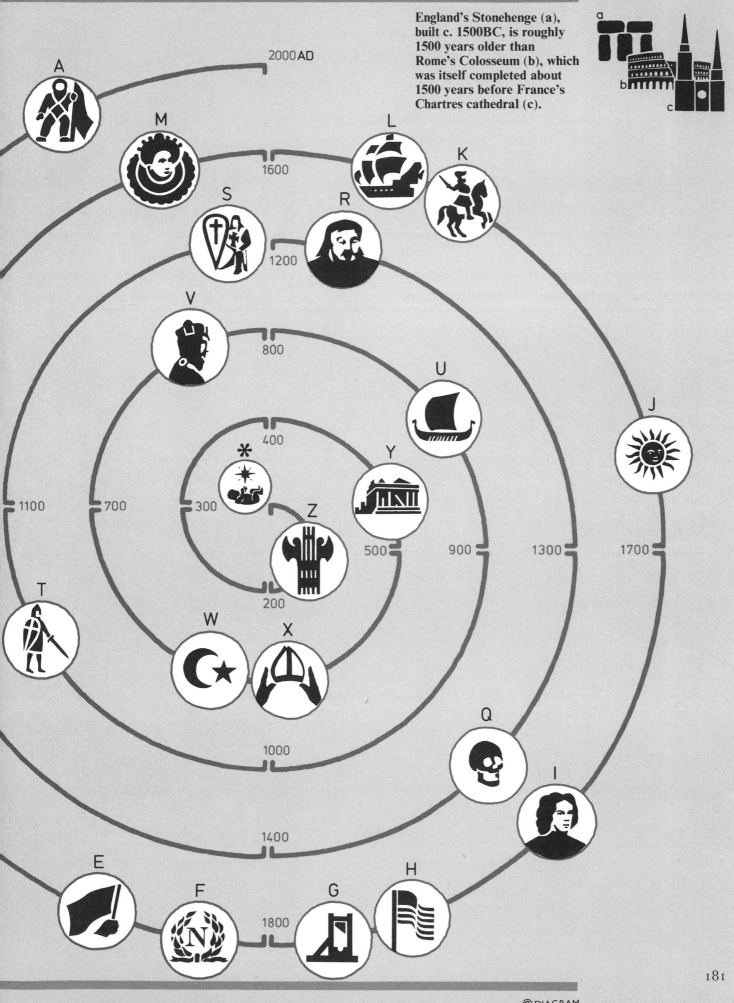

England's Stonehenge (a), built c. 1500BC, is roughly 1500 years older than Rome's Colosseum (b), which was itself completed about 1500 years before France's Chartres cathedral (c).

181

CHAPTER 9

Like father, like son—
Sir Malcolm Campbell and
his son Donald are here
photographed in 1933 with
the record-breaking car
Bluebird. Campbell senior
raised the world land-
speed record nine times in
1924–35. His son took the
land-speed record once, in
1964, and the water-speed
record seven times before
his death in 1967 (BBC
Hulton Picture Library).

SPEED

The running action of a dog is shown in this sequence of photographs from Eadweard Muybridge's *Animal Locomotion*, published in 1887 (Kingston-upon-Thames Museum and Art Gallery).

Flying through the air on a cannonball—just one of the fabulous exploits described in R. E. Raspe's *Baron Münchhausen's Narrative of His Marvellous Travels and Campaigns in Russia*, first published in 1785 (The Mansell Collection).

MEASURING SPEED

In this chapter we look at comparisons of speed, concentrating on its definition as the amount of time taken to travel a particular distance. Speed is calculated by dividing a distance measurement by a time measurement, and is expressed in units that combine both these factors, as for example in kilometers per hour.

1 mile per hour (mph)	=	1.6093 kilometers per hour
1 yard per minute (yd/min)	=	0.9144 meters per minute
1 foot per second (ft/s)	=	0.3048 meters per minute
1 inch per second (in/s)	=	2.5400 centimeters per second
1 kilometer per hour (km/h)	=	0.6214 miles per hour
1 meter per minute (m/min)	=	1.0936 yards per minute
1 meter per minute (m/min)	=	3.2808 feet per minute
1 centimeter per second (cm/s)	=	0.3937 inches per second
1 mile per hour	=	1.4667 feet per second
1 kilometer per hour	=	0.2778 meters per second

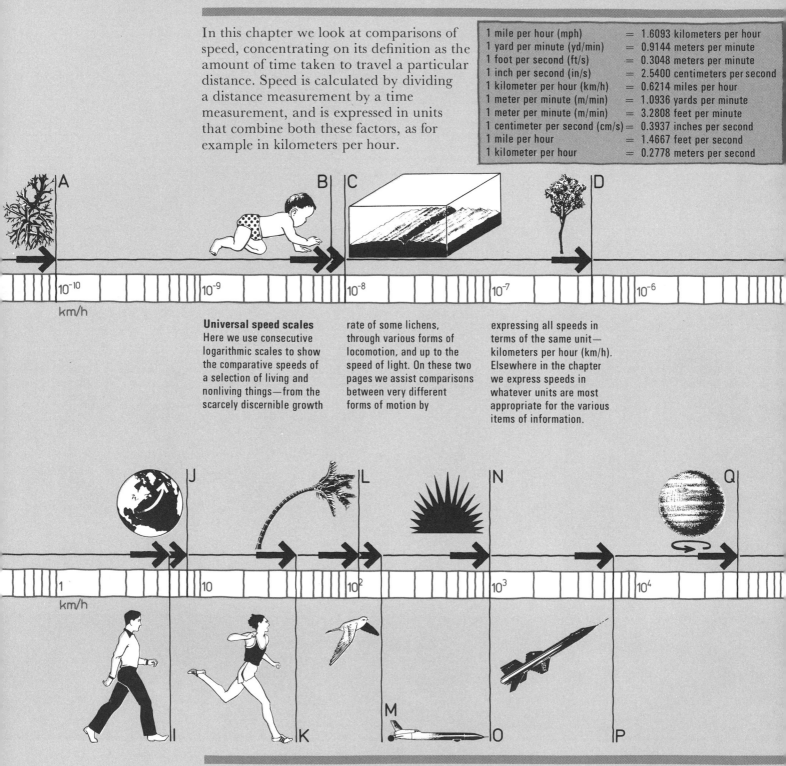

Universal speed scales
Here we use consecutive logarithmic scales to show the comparative speeds of a selection of living and nonliving things—from the scarcely discernible growth rate of some lichens, through various forms of locomotion, and up to the speed of light. On these two pages we assist comparisons between very different forms of motion by expressing all speeds in terms of the same unit— kilometers per hour (km/h). Elsewhere in the chapter we express speeds in whatever units are most appropriate for the various items of information.

Earth has a rotational velocity at the Equator of 1040mph (1674km/h). This means that anyone standing on the Equator is traveling some 1.7 times faster than the official world land-speed record.

The orbital speed of an electron in a uranium atom (8,700,000km/h) is fifty times faster than the mean orbital velocity of Mercury around the Sun (172,404km/h).

The Gulf Stream current's average flow of 5.2mph (8.4km/h) is more than 18,000 times faster than an Antarctic glacier's average flow of 84.6yd (77.4m) a week.

Conversion of units *left*
Listed are abbreviations and conversion factors for common units of speed. In addition to metric equivalents for US/imperial measurements and vice versa, we give the commonly needed conversions of mph into ft/s, and km/h into m/s.

Knots for sailors
Nautical speeds are given in knots (nautical miles per hour). There are two units: international knots (kn) and British (UKkn).
1kn = 1.1508mph
1kn = 1.8520km/h
1UKkn = 1.1515mph
1UKkn = 1.8531km/h

Super Mach speeds *right*
The Mach scale is used to express speeds faster than sound, with Mach 1 equal to the speed of sound, Mach 2 to twice the speed of sound, and so on. But the speed of sound is not a constant, since it depends on the substance through which the sound waves are passing, and on temperature. For land speeds (sea level, 15°C) Mach 1 is 760.98mph (1224.65km/h). Sound travels more slowly in the cooler stratosphere, and for air speeds Mach 1 is taken as 659.78mph (1061.78km/h).

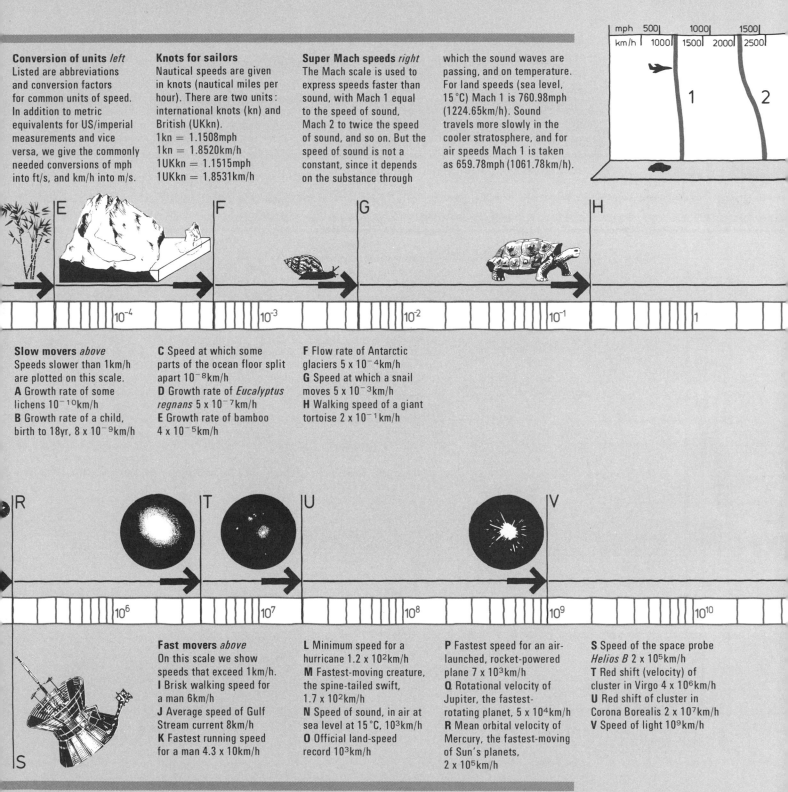

Slow movers *above*
Speeds slower than 1km/h are plotted on this scale.
A Growth rate of some lichens 10^{-10}km/h
B Growth rate of a child, birth to 18yr, 8×10^{-9}km/h

C Speed at which some parts of the ocean floor split apart 10^{-8}km/h
D Growth rate of *Eucalyptus regnans* 5×10^{-7}km/h
E Growth rate of bamboo 4×10^{-5}km/h

F Flow rate of Antarctic glaciers 5×10^{-4}km/h
G Speed at which a snail moves 5×10^{-3}km/h
H Walking speed of a giant tortoise 2×10^{-1}km/h

Fast movers *above*
On this scale we show speeds that exceed 1km/h.
I Brisk walking speed for a man 6km/h
J Average speed of Gulf Stream current 8km/h
K Fastest running speed for a man 4.3×10km/h

L Minimum speed for a hurricane 1.2×10^{2}km/h
M Fastest-moving creature, the spine-tailed swift, 1.7×10^{2}km/h
N Speed of sound, in air at sea level at 15°C, 10^{3}km/h
O Official land-speed record 10^{3}km/h

P Fastest speed for an air-launched, rocket-powered plane 7×10^{3}km/h
Q Rotational velocity of Jupiter, the fastest-rotating planet, 5×10^{4}km/h
R Mean orbital velocity of Mercury, the fastest-moving of Sun's planets, 2×10^{5}km/h

S Speed of the space probe *Helios B* 2×10^{5}km/h
T Red shift (velocity) of cluster in Virgo 4×10^{6}km/h
U Red shift of cluster in Corona Borealis 2×10^{7}km/h
V Speed of light 10^{9}km/h

It is not surprising that we see lightning before we hear thunder, for light travels at 186,282 miles per second (roughly 10^{9} km/h)—more than 880,000 times faster than the 760.98 miles per hour (roughly 10^{3} km/h) that sound travels through air at sea level at 15°C.

SPEEDS IN NATURE

All things in nature—nonliving as well as living—are constantly on the move. Even the rock rafts that make up Earth's crust shift up and down and to and fro—but generally only by a fraction of an inch per year. Among living things change is more obvious. Plant growth, for example, can be accurately timed and measured.

Plant growth *left*
Shown here to scale are daily growth rates for some fast- and slow-growing plants. A bamboo can grow as many inches in one day as an average child grows in the first 10 years after birth. At the slow end of the plant growth spectrum are some types of lichen that take a century to grow just one inch.
a Bamboo 35.4in
b Certain seaweeds 17.7in
c Bermuda grass 5.9in
d *Albizzia falcatoria* 1.1in
e *Eucalyptus regnans* 0.5in
f Sitka spruce growing on polar tree line 0.0003in
g Some lichens 0.0001in

Beaufort scale
The diagram *below* and the table *right* give Beaufort scale wind force numbers together with the range of wind speeds to which each number is applied. (Note that speeds in mph and km/h are not exact equivalents.) Also included in the table are official descriptive titles for the Beaufort scale forces, and examples of physical manifestations that characterize them. The latter are indicated on the diagram by symbols.

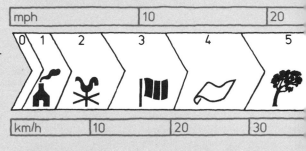

Rivers of ice *right*
A glacier is a frozen river that flows slowly down a valley. A body frozen into a glacier high up in its valley may eventually emerge perfectly preserved lower in the valley where the glacier is melting. Glaciers in different places flow at different rates. Shown here beneath a 66yd-long ice-hockey pitch are the distances traveled in one week by the ice in three types of glacier from three parts of the world.
A Alpine glacier 304.6yd
B Greenland glacier 236.9yd
C Antarctic glacier 84.6yd

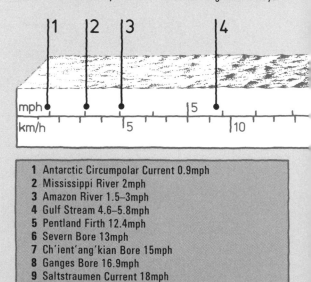

1 Antarctic Circumpolar Current 0.9mph
2 Mississippi River 2mph
3 Amazon River 1.5–3mph
4 Gulf Stream 4.6–5.8mph
5 Pentland Firth 12.4mph
6 Severn Bore 13mph
7 Ch'ient'ang'kian Bore 15mph
8 Ganges Bore 16.9mph
9 Saltstraumen Current 18mph
10 Lava Falls, Colorado River 30mph

Parts of the ocean floor split apart at the rate of about four inches a year. So two mermaids seated side by side on either side of a split could be at arm's length after fifteen years!

A wind as fast as the fastest speed run by a man (27mph) is classed as a "strong breeze" on the Beaufort scale. A wind as fast as a running cheetah (70mph) is classed as a "storm."

Number	Description	Speed		Characteristics
0	Calm	<1mph	<1km/h	Smoke rises vertically
1	Light air	1–3mph	1–5km/h	Direction shown by smoke
2	Light breeze	4–7mph	6–12km/h	Direction shown by wind vane
3	Gentle breeze	8–12mph	13–20km/h	Wind extends a light flag
4	Moderate breeze	13–18mph	21–29km/h	Raises dust and loose paper
5	Fresh breeze	19–24mph	30–39km/h	Small trees in leaf start to sway
6	Strong breeze	25–31mph	40–50km/h	Umbrellas used with difficulty
7	Moderate gale	32–38mph	51–61km/h	Inconvenient to walk against wind
8	Fresh gale	39–46mph	62–74km/h	Twigs broken off trees
9	Strong gale	47–54mph	75–87km/h	Chimney pots and slates removed
10	Whole gale	55–63mph	88–102km/h	Trees uprooted; considerable damage
11	Storm	64–75mph	103–120km/h	Widespread damage
12–17	Hurricane	>75mph	>120km/h	Extremely violent

Windy mountain
Mount Washington in the USA has experienced winds of up to 231mph (371km/h)—some 95mph (153km/h) beyond Beaufort scale force 17.

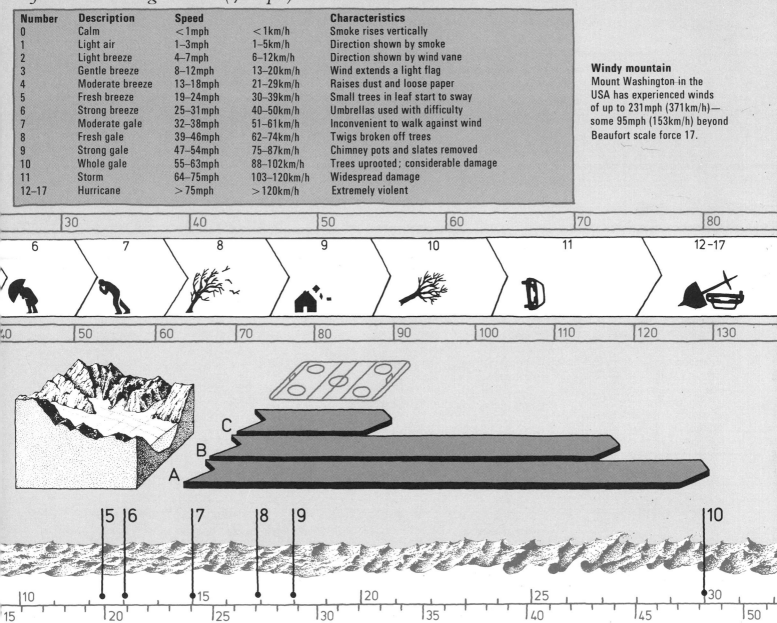

Water speeds
Listed *left* and shown *above* are the speeds that water moves in selected ocean currents, rivers, and bores (tidal waves that surge up rivers from narrow estuaries). Ocean-current speeds vary enormously. Thus Norway's Saltstraumen Current is 20 times faster than part of the Antarctic Circumpolar Current. River flow varies with the slope of the bed. The speed of normal ocean waves depends largely on the winds that generate them. Many waves move through the sea at speeds of 18–56mph.

Tsunamis
Often misnamed tidal waves, tsunamis are giant, high-speed waves unleashed by sub-oceanic earthquakes. A tsunami can cross an ocean at up to 490mph. This is 8mph faster than the official speed record for a piston-engined plane.

The tidal bore of England's River Severn has reached a top speed of 13mph, very slightly slower than the top speed recorded for a rowing eight over a 2000-meter still-water course (13.46mph).

©DIAGRAM

ANIMALS AND MAN

When it comes to natural movement many animals are faster than man, although man is by no means one of nature's slowcoaches. Just as man's inventions enable him to travel faster in the air than on land or in water, it is in the air that the highest of all animal speeds are recorded. The spine-tailed swift is at least half as fast again as the cheetah. The speeds given here have been standardized as far as possible, but when dealing with wild animals there is an almost inevitable lack of standard methods for timing and for measuring distances and wind speeds. Another complication is that exceptional individuals are sometimes taken to be representative of an entire species.

Speeds in the air
right The racing pigeon would come a poor second in a race with the fastest of all birds, the spine-tailed swift. Selected insects are also included for comparison; small size makes their performances particularly impressive.

Spine-tailed swift 106.25mph
Pigeon 60mph
Hawk moth 33mph
Monarch butterfly 20mph
Honeybee 11mph

Speeds on land
right The cheetah is the fastest animal on land, but only for distances up to about 350yd, after which the prong-horned antelope is faster. Man's fastest known speed was calculated from the top speed touched during a 100yd race.

Cheetah 70mph
Prong-horned antelope 60mph
Jackrabbit 45mph
Ostrich 30 + mph
Man 27mph

Speeds in water
right The sailfish is thought to be the fastest creature in water, but data on aquatic creatures is extremely difficult to verify. The speed of 5.19mph for man is calculated from the top speed touched in a sprint.

Sailfish 60 + mph
Flying fish 40 + mph
Dolphin 37mph
Trout 15mph
Man 5.19mph

10   20
10   20   30   40

Slow and not so slow
right Here we compare the time taken by some of nature's slower creatures to travel 100m, which the world record holder J.R. Hines covered in 9.95sec and a typical young adult walks in about 1min.
a Mole 7hr 50min—3hr faster than a mechanical mole can burrow the same distance!
b Snail 2hr 4min
c Giant tortoise 22min
d Three-toed sloth 22min
e Spider 8min 50sec
f Centipede 3min 25sec—quite an achievement in view of its small size!

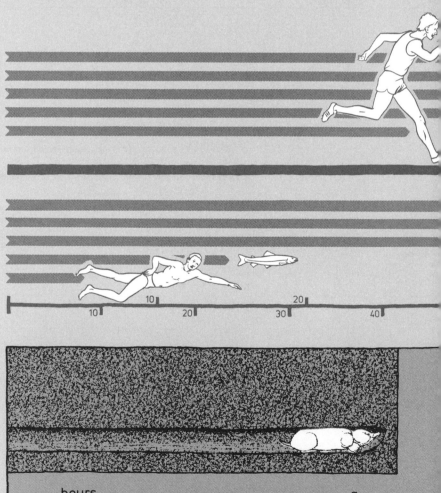

hours

a

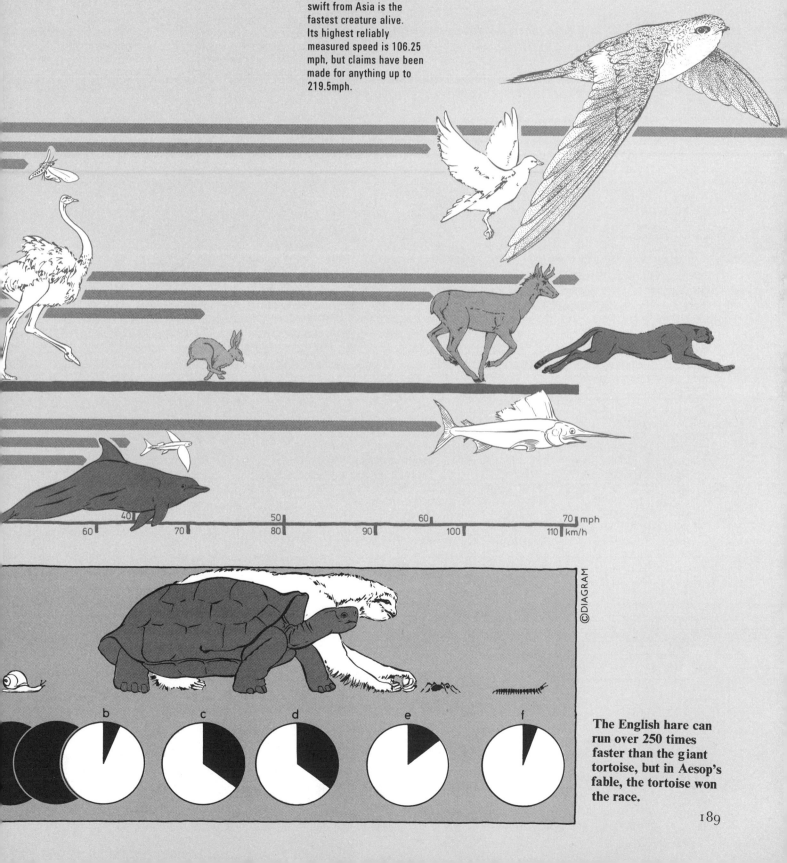

From a stationary start, man could hold the lead over a modern sports car for approximately 10 yd.

A close bet—over 110yd, the fastest timed greyhound (41.72mph for 410yd) would have lost by 3-4yd to the fastest horse (43.26mph for a 440yd race).

Speed

Fastest creature
right The spine-tailed swift from Asia is the fastest creature alive. Its highest reliably measured speed is 106.25 mph, but claims have been made for anything up to 219.5mph.

| 40 | 50 | 60 | 70 mph |
| 60 | 70 | 80 | 90 | 100 | 110 km/h |

©DIAGRAM

b   c   d   e   f

The English hare can run over 250 times faster than the giant tortoise, but in Aesop's fable, the tortoise won the race.

HUMAN ACHIEVEMENTS

Man's muscular machinery has probably remained essentially the same for many thousands of years. Yet each year athletes continue to improve on existing world speed records. Most of the speeds given here would have seemed impossible in 1900, but few if any of them are now expected to be records in the year 2000.

Sporting speeds
below The diagram shows top speeds achieved by athletes in different speed events. Muscles or gravity provided the motive force for all these achievements, although the cyclist was aided by the slipstream of the vehicle ahead of him.

1 Roller skating, 25.8mph over 440yd of road.
2 Running, 27mph+ attained during 100yd dash.
3 Speed skating, 30.22mph averaged over 500m.
4 Skateboarding, 53.45mph in stand-up position.
5 Luge tobogganing, 80mph+ recorded.

Event		Holder	Time	Speed
A 100m	Men	J. R. Hines (USA)	9.95s	10.05m/s
	Women	M. Göhr (GDR)	10.88s	9.19m/s
B 200m	Men	P. Mennea (Italy)	19.72s	10.14m/s
	Women	M. Koch (GDR)	21.71s	9.21m/s
C 400m	Men	L. E. Evans (USA)	43.86s	9.12m/s
	Women	M. Koch (GDR)	48.60s	8.23m/s
D 800m	Men	S. Coe (UK)	1min 42.4s	7.81m/s
	Women	T. Kazankina (USSR)	1min 54.9s	6.96m/s
E 1500m	Men	S. Coe (UK)	3min 32.1s	7.07m/s
	Women	T. Kazankina (USSR)	3min 56.0s	6.36m/s
F 5000m	Men	H. Rono (Kenya)	13min 8.4s	6.34m/s
G 10,000m	Men	H. Rono (Kenya)	27min 22.4s	6.09m/s

Track event records
Listed *left* are details of world records for 12 track events, as at April 1, 1980. Each of the speeds given— also shown in the diagram *below*—is the number of meters run per second as averaged over the whole distance of the race.

Olympic runners achieve faster average speeds over 200m than they do over 100m. This is because a comparatively slow speed for the first few meters of both these events is averaged out in the 200m by a longer distance run at full speed.

The fastest speed attained by a water skier (134.3mph) is almost exactly 10mph faster than the fastest speed touched by a skier on land (124.4mph).

6 Cresta run tobogganing, 90mph sometimes reached.
7 Skibob, top speed of 103.4mph attained.
8 Skiing, 124.4mph reached on a downhill course.
9 Cycling, average speed of of 140.5mph over ¾mi, attained behind a windshield mounted on an automobile.

Feats of leg power *right*
The fastest speed touched by a runner, probably in excess of 27mph (**a**), is here compared with the top speed of 49.38mph attained by a cyclist without the help of a windshield (**b**).

©DIAGRAM

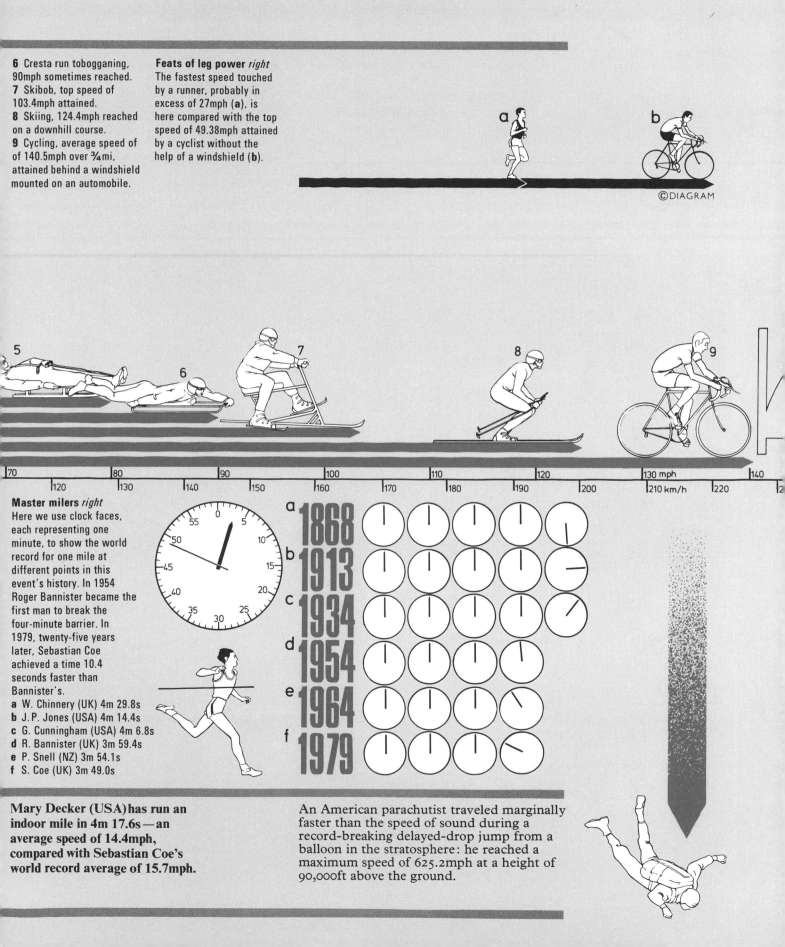

Master milers *right*
Here we use clock faces, each representing one minute, to show the world record for one mile at different points in this event's history. In 1954 Roger Bannister became the first man to break the four-minute barrier. In 1979, twenty-five years later, Sebastian Coe achieved a time 10.4 seconds faster than Bannister's.
a W. Chinnery (UK) 4m 29.8s
b J. P. Jones (USA) 4m 14.4s
c G. Cunningham (USA) 4m 6.8s
d R. Bannister (UK) 3m 59.4s
e P. Snell (NZ) 3m 54.1s
f S. Coe (UK) 3m 49.0s

a 1868
b 1913
c 1934
d 1954
e 1964
f 1979

Mary Decker (USA) has run an indoor mile in 4m 17.6s — an average speed of 14.4mph, compared with Sebastian Coe's world record average of 15.7mph.

An American parachutist traveled marginally faster than the speed of sound during a record-breaking delayed-drop jump from a balloon in the stratosphere: he reached a maximum speed of 625.2mph at a height of 90,000ft above the ground.

SPEEDS ON LAND

Land-speed records fell repeatedly as man exploited more effective kinds of motive force. In 1907 the steam-powered *Stanley Steamer* reached 150mph (241km/h), the then highest speed for a manned vehicle in any element. Since 1959 the fastest land vehicle has been a rail-mounted, unmanned rocket sled four times as fast as sound.

High-speed cars
The table *right* lists a selection of official land-speed record holders. (From 1911 only the average speed of two runs, in opposite directions, has qualified for official recognition.) The diagram *below* plots selected record-holders on a graph showing how record speeds increased as the years passed and as direct thrust replaced propulsion via the wheels. A wheel-driven vehicle reached 409.27mph in 1965. That year a jet car reached 600.60mph. Since 1970 rocket cars have led the field.

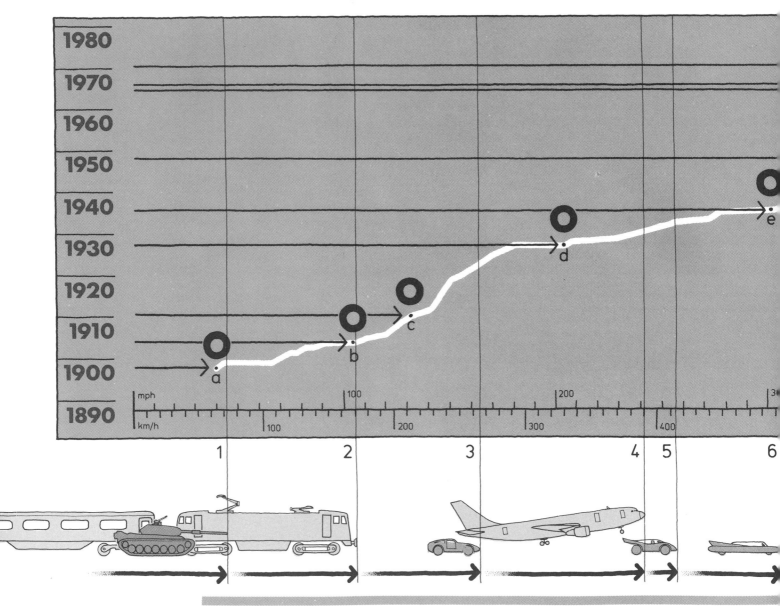

Fastest speeds (not official records) attained in different types of vehicle are:
418.504mph for a piston-engined car (R. S. Summers, *Goldenrod*, 1965)
429.311mph for a wheel-driven car (D. Campbell, *Proteus Bluebird*, 1964)
613.995mph for a jet-engined car (N. C. Breedlove, *Spirit of America—Sonic 1*, 1965)
739.67mph for a rocket-engined car (S. Barret, *Budweiser Rocket*, 1979)

To match the distance covered in one hour by the US Air Force rocket sled (3090 miles), the fastest-ever runner would have to maintain his top speed of 27mph from midday on Monday until 6am Saturday.

Holder (year of record)	Vehicle	Speed	
a G. de Chasseloup-Laubat (1898)	Jeantaud	39.24mph	63.15km/h
b P. Baras (1904)	Darracq	104.52mph	168.20km/h
c B. Oldfield (1910)	Blitzen Benz	131.72mph	211.98km/h
d H. O. D. Segrave (1927)	Sunbeam	203.79mph	327.96km/h
e M. Campbell (1935)	Campbell Special	301.13mph	484.61km/h
f J. R. Cobb (1947)	Railton	394.20mph	634.39km/h
g A. Arfons (1964)	*Green Monster*	536.71mph	863.73km/h
h N. C. Breedlove (1965)	*Spirit of America—Sonic I*	600.60mph	966.55km/h
i G. Gabelich (1970)	*The Blue Flame*	622.29mph	1001.45km/h

First and last *below*
In 1898 G. de Chasseloup-Laubat in a Jeantaud (**a**) set a land-speed record of 39.24mph. The present record of 622.29mph, set in 1970 by G. Gabelich in *The Blue Flame* (**i**) is 15.9 times faster.

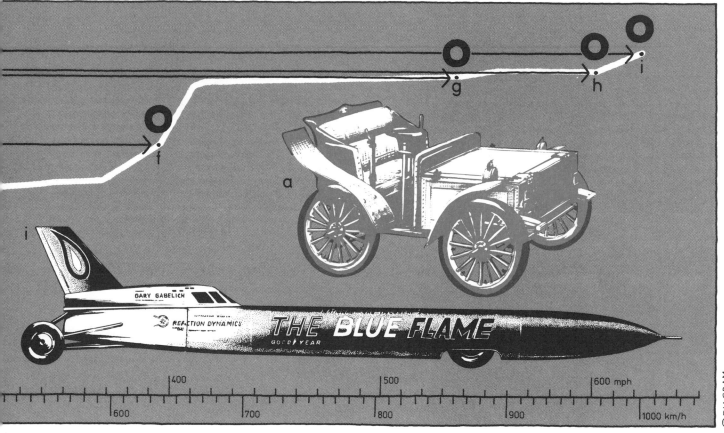

©DIAGRAM

Hares and tortoises *left*
Plotted against the same scale as the graph of official land-speed records are the top land speeds of six vehicles, most far slower than recent land-speed record holders. (Only in the air do travelers routinely approach the record speed on land.) Listed in order of their speeds these vehicles are:
1 Tank 45mph
2 Regular electric passenger train 106.25mph
3 Production car 163mph
4 Aircraft (landing) 242mph
5 Racing car 257mph
6 Motorcycle 307.69mph

Rocket on rails *below*
The highest speed so far attained on land is 3090mph, by an unmanned US Air Force rocket-powered sled on a railed track at Holloman, New Mexico in 1959. This speed is almost five times faster than the regular land speed record.

SPEEDS ON WATER

The official world water-speed record is only just over half as fast as its land equivalent, and approximately one-seventh as fast as that in air. It is, however, the most dangerous event of the three—for a boat easily goes out of control at high speeds. Also included here for comparison are various other speed records on water.

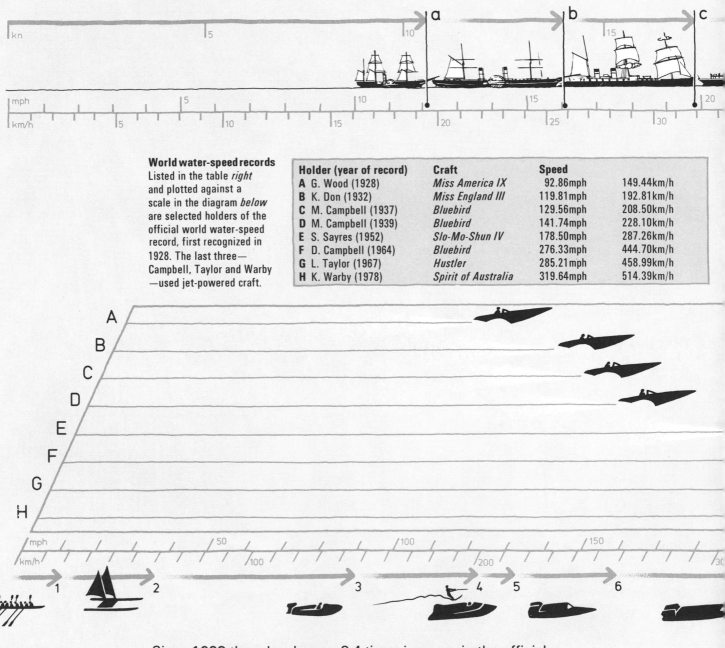

World water-speed records Listed in the table *right* and plotted against a scale in the diagram *below* are selected holders of the official world water-speed record, first recognized in 1928. The last three—Campbell, Taylor and Warby —used jet-powered craft.

Holder (year of record)	Craft	Speed	
A G. Wood (1928)	*Miss America IX*	92.86mph	149.44km/h
B K. Don (1932)	*Miss England III*	119.81mph	192.81km/h
C M. Campbell (1937)	*Bluebird*	129.56mph	208.50km/h
D M. Campbell (1939)	*Bluebird*	141.74mph	228.10km/h
E S. Sayres (1952)	*Slo-Mo-Shun IV*	178.50mph	287.26km/h
F D. Campbell (1964)	*Bluebird*	276.33mph	444.70km/h
G L. Taylor (1967)	*Hustler*	285.21mph	458.99km/h
H K. Warby (1978)	*Spirit of Australia*	319.64mph	514.39km/h

Since **1928** there has been a 3.4 times increase in the official water-speed record. Over the same period there has been a 3 times increase in the official land-speed record, and a 6.9 times increase in the official air-speed record.

In 1952 the liner *United States* took the Blue Riband award for the fastest Atlantic crossing. Her average speed of 35.6 knots for an eastward crossing was only 3.4 times faster than the average speed for a similar crossing by *Britannia*, which won the award 112 years earlier.

A sailfish can swim half as fast again as the average speed maintained on the *United States'* Blue Riband run.

Speed

Liner (year of record)	Average speed		
a *Britannia* (1840)	10.6kn	12.2mph	19.6km/h
b *Scotia* (1863)	14.0kn	16.1mph	25.9km/h
c *Alaska* (1882)	17.2kn	19.8mph	31.9km/h
d *Kaiser Wilhelm der Grosse* (1897)	22.4kn	25.8mph	41.5km/h
e *Mauretania* (1909)	25.9kn	29.8mph	48.0km/h
f *Bremen* (1929)	27.9kn	32.1mph	51.7km/h
g *Queen Mary* (1938)	31.7kn	36.5mph	58.7km/h
h *United States* (1952)	35.6kn	41.0mph	66.0km/h

Blue Riband speeds
Listed *left* and illustrated and plotted against the scale *below* is a selection of liners that made record transatlantic crossings from the United States to Europe to win the coveted Blue Riband award. Average speeds are given in international knots (see p. 185) and in miles and kilometers per hour. Two ships have held the record continuously for periods of 14 years: *Mauretania* from 1909–23, and *Queen Mary* from 1938–52. *United States* has now held the record for 28 years.

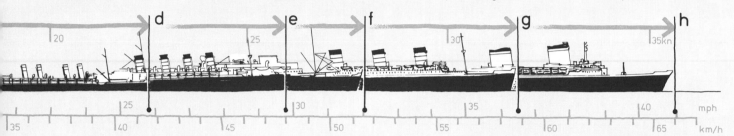

Holder (year of record)	Craft	Record	Speed	
1 East German team (1976)		Rowing eight (2000m, still-water)	13.46mph	21.66km/h
2 T. Colman (1977)	*Crossbow II*	Sailing speed	38.46mph	61.89km/h
3 M. Frode (1977)	Class IIID Frode	Off-shore powerboat	92.99mph	149.65km/h
4 D. Churchill (1971)		Water-skiing speed	125.69mph	202.27km/h
5 J. F. Merton (1973)	*Quicksilver*	Class ON powerboat	136.38mph	219.48km/h
6 S. Williams (1977)	U-96 KYYX	Women's water-speed	163.04mph	262.38km/h
7 L. Hill (1971)	*Mr Ed*	Propellor-driven	202.42mph	325.75km/h

Other records on water
Described *left* and plotted *bottom* are official world records (April 1980) for selected water events. Conditions of recognition obviously vary from event to event, but each record listed is approved by its own international federation.

Fastest speed on water
The fastest speed ever attained on water is an estimated 345mph (556km/h) by K. Warby in 1978 in his *Spirit of Australia* (**H**). This is approximately 8% faster than his official water-speed record.

SPEEDS IN THE AIR

In the 73 years since Orville Wright's first controlled powered flight in 1903 official air-speed records have risen by an annual average equal to the speed of Wright's flight (30mph), for in 1976 an orthodox jet plane flew 73 times faster than Wright. Air-launched planes propelled by rockets are dramatically faster still.

Air speed records
The table *right* and the diagram *below* show details of official air-speed records spanning more than six decades. The speeds shown were all ratified by the Fédération Aéronautique Internationale. They exclude records set by air-launched aircraft. Vertical lines indicate Mach speeds: multiples of the speed of sound, taken as 659.78mph (1061.78km/h) in the stratosphere (see p. 185). From 1909 official record speeds doubled or more than doubled at intervals of 1, 10, 8, 24 and 6 years.

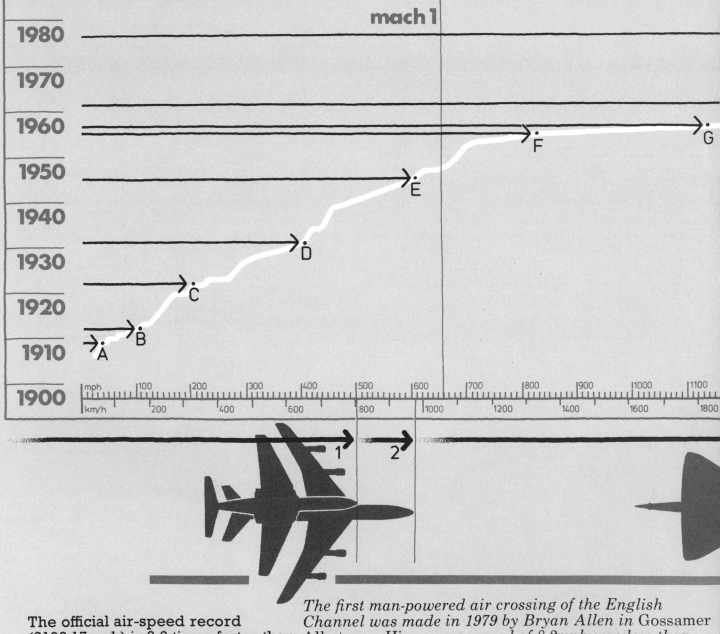

The official air-speed record (2193.17mph) is 3.6 times faster than the cruising speed of a Boeing 707 (604mph) and 1.5 times faster than that of *Concorde* (1450mph).

The first man-powered air crossing of the English Channel was made in 1979 by Bryan Allen in Gossamer Albatross. *His average speed of 8.2mph was less than one-fifth that of Louis Blériot, who in 1909 made the first air crossing of the Channel in his* Blériot XI *monoplane at an average speed of 42.7mph.*

The holder of the official air-speed record flew at a speed of 2193.17mph, or 3216.72 feet per second. The speed of a bullet from the US Army's current M16 rifle is even faster—3300ft per second.

Holder (year of record)		Aircraft	Speed	
A	P. Tissandier (1909)	Wright biplane	34.03mph	54.76km/h
B	J. Védrines (1912)	Deperdussin monoplane	100.21mph	161.27km/h
C	S. Lecointe (1922)	Nieuport-Delage 29	205.20mph	330.23km/h
D	G.H. Stainforth (1931)	Supermarine S6B	406.94mph	654.89km/h
E	H.J. Wilson (1945)	Gloster Meteor F4	606.25mph	975.64km/h
F	H.A. Hanes (1955)	F-100C Super Sabre	822.09mph	1322.99km/h
G	P. Twiss (1956)	Fairey Delta 2	1131.76mph	1820.12km/h
H	R.B. Robinson (1961)	McDonnell F4H-1F Phantom II	1606.51mph	2585.36km/h
I	E.W. Joersz & G.T. Morgan (1976)	Lockheed SR-71A	2193.17mph	3529.47km/h

First supersonic flight
above A Bell XS-1 rocket plane was the first to fly faster than sound. In 1947 USAF Captain Charles E. Yeager flew at 670mph (1078km/h) at 42,000ft (12,800m). No land vehicle matched his speed for the next 32 years.

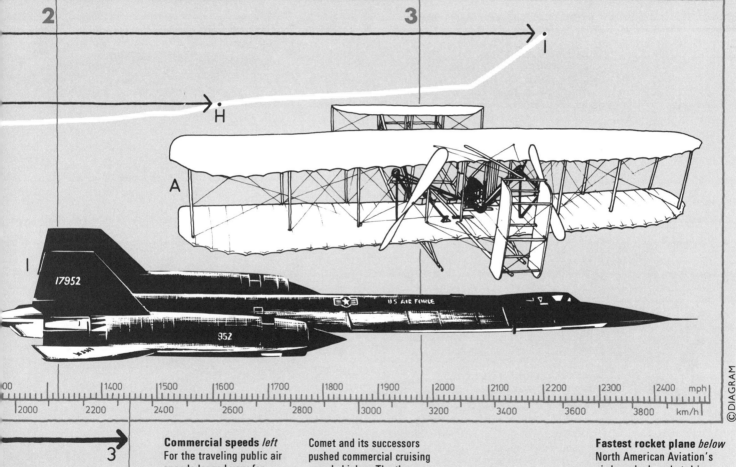

©DIAGRAM

Commercial speeds *left*
For the traveling public air speeds lagged very far behind world air-speed records. By the late 1930s US passenger aircraft averaged 158mph (254km/h). By the early 1950s speeds reached 370mph (595km/h). Then the turbojet-powered Comet and its successors pushed commercial cruising speeds higher. The three planes *left* span the range of cruising speeds for airliners, *Concorde* almost trebling the Comet's speed.
1 Comet 4, 500mph
2 Boeing 707, 604mph
3 *Concorde*, 1450mph

Fastest rocket plane *below*
North American Aviation's air-launched, rocket-driven X-15A-2 attained a speed of 4534mph (7297km/h) in 1967. This was more than twice as fast as the official air-speed record of 2193.17mph (3529.47km/h) set some 9 years later.

VEHICLE SPEEDS

In the early 1800s steam locomotives amazed spectators by moving as quickly as a horse could gallop. Today, vehicles on land and water and in air and space attain speeds undreamed of a lifetime ago. The speeds given here are chiefly top speeds touched, not official records set in specified conditions.

Subsonic records
Plotted on the scale *below* and listed in the table *right* are the fastest speeds achieved by certain types of vehicle on land, water, ice, snow and in air. The list excludes vehicles capable of producing supersonic performances.

Limitations of certain propulsive methods mean some of these speeds may be never more than marginally bettered. Thus the steam locomotive record set in 1938 still stood in 1980. But by then rail speeds were being doubled by the linear induction motor.

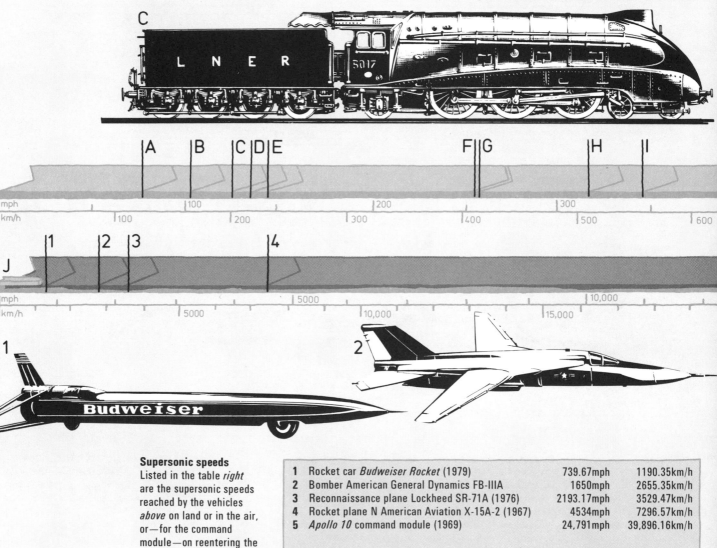

Supersonic speeds
Listed in the table *right* are the supersonic speeds reached by the vehicles *above* on land or in the air, or—for the command module—on reentering the atmosphere from space. All of these supersonic vehicles were manned.

		mph	km/h
1	Rocket car *Budweiser Rocket* (1979)	739.67mph	1190.35km/h
2	Bomber American General Dynamics FB-IIIA	1650mph	2655.35km/h
3	Reconnaissance plane Lockheed SR-71A (1976)	2193.17mph	3529.47km/h
4	Rocket plane N American Aviation X-15A-2 (1967)	4534mph	7296.57km/h
5	*Apollo 10* command module (1969)	24,791mph	39,896.16km/h

In 1979 S. Barret in the rocket-powered car *Budweiser Rocket* touched a speed of 739.67mph. This is approximately 55 times faster than the 13.5mph reached in 1830 by George Stephenson's famous *Rocket* steam locomotive.

The jet-powered tracked hovercraft L'Aérotrain 02 *has touched a speed of 255.3mph. This is three times faster than the average speed of 85mph for the fastest daily run of a British Rail train from London to Edinburgh.*

A	Sand yacht (1975)	77.47mph	124.67km/h
B	Military hovercraft SES-100B (1976)	102.35mph	164.71km/h
C	Steam locomotive *Mallard* (1938)	126mph	202.77km/h
D	Snowmobile (1977)	135.93mph	218.75km/h
E	Ice yacht (1938)	143mph	230.13km/h
F	LIMRV rail research vehicle (1974)	254.76mph	409.99km/h
G	Monorail hovercraft *L'Aérotrain 02* (1968)	255.3mph	410.85km/h
H	Helicopter Bell YUH-1B (1969)	316.1mph	508.70km/h
I	Hydroplane *Spirit of Australia* (1977)	345.23mph	555.58km/h
J	Rocket car *The Blue Flame* (1970)	650mph	1046.05km/h

Speeds in space
below This diagram compares the top speed for a manned space vehicle (the *Apollo 10* command module) with the speeds attained by two robot space probes.
a *Apollo 10* (1969) attained 24,791mph (39,896km/h) on reentering the atmosphere.

b *Pioneer 10* (1972) reached an escape velocity of 32,114mph (51,681km/h) and became the first man-made craft to travel fast enough to leave the solar system.
c *Helios B* (1976) reached 149,125mph (239,987km/h) in space flight on its mission as a solar probe.

©DIAGRAM

The solar probe *Helios B* **reached a speed of 149,125mph. Traveling at this speed for its entire journey it would have crossed the orbit of the Moon after about 1½ hours and reached the Sun in 26 days.**

THE PLANETS

Although we cannot feel it, Earth is moving rapidly through space. Our planet's mean orbital velocity—the average speed at which it orbits the Sun—is some 66,641mph. Its rotational velocity at the Equator—the speed it turns on its axis—is 1040mph. Here we compare Earth's speeds with those of other planets.

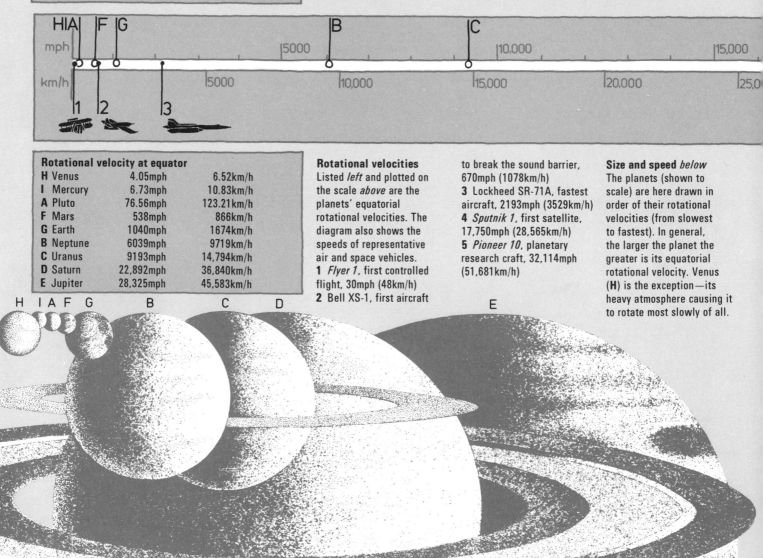

Mean orbital velocity		
A Pluto	10,604mph	17,064km/h
B Neptune	12,147mph	19,548km/h
C Uranus	15,234mph	24,516km/h
D Saturn	21,565mph	34,704km/h
E Jupiter	29,216mph	47,016km/h
F Mars	53,980mph	86,868km/h
G Earth	66,641mph	107,244km/h
H Venus	78,364mph	126,108km/h
I Mercury	107,132mph	172,404km/h

Orbital velocities *left*
Listed are the planets' mean orbital velocities, from slowest to fastest. The nearer a planet is to the Sun the faster it has to travel to counteract the Sun's gravitational force that would otherwise pull the planet toward it.

Orbital velocity scale *above* In this diagram the mean orbital velocities of the planets are shown against an mph and km/h scale. The speed for Mercury, with the highest orbital velocity, is about 10 times faster than that for Pluto, with the lowest.

Race to the Moon *right* Here the orbital velocities of the planets are compared by marking the distance traveled by each planet in one hour on a line representing the center-to-center distance between Earth and Moon, 238,840mi (384,365km).

Rotational velocity at equator		
H Venus	4.05mph	6.52km/h
I Mercury	6.73mph	10.83km/h
A Pluto	76.56mph	123.21km/h
F Mars	538mph	866km/h
G Earth	1040mph	1674km/h
B Neptune	6039mph	9719km/h
C Uranus	9193mph	14,794km/h
D Saturn	22,892mph	36,840km/h
E Jupiter	28,325mph	45,583km/h

Rotational velocities
Listed *left* and plotted on the scale *above* are the planets' equatorial rotational velocities. The diagram also shows the speeds of representative air and space vehicles.
1 *Flyer 1*, first controlled flight, 30mph (48km/h)
2 *Bell XS-1*, first aircraft to break the sound barrier, 670mph (1078km/h)
3 Lockheed SR-71A, fastest aircraft, 2193mph (3529km/h)
4 *Sputnik 1*, first satellite, 17,750mph (28,565km/h)
5 *Pioneer 10*, planetary research craft, 32,114mph (51,681km/h)

Size and speed *below*
The planets (shown to scale) are here drawn in order of their rotational velocities (from slowest to fastest). In general, the larger the planet the greater is its equatorial rotational velocity. Venus (**H**) is the exception—its heavy atmosphere causing it to rotate most slowly of all.

Mercury has the fastest mean orbital velocity but the second slowest equatorial rotational velocity. Jupiter has the fastest equatorial rotational velocity while ranking fifth for mean orbital velocity.

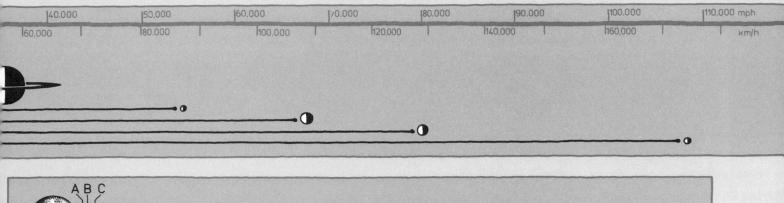

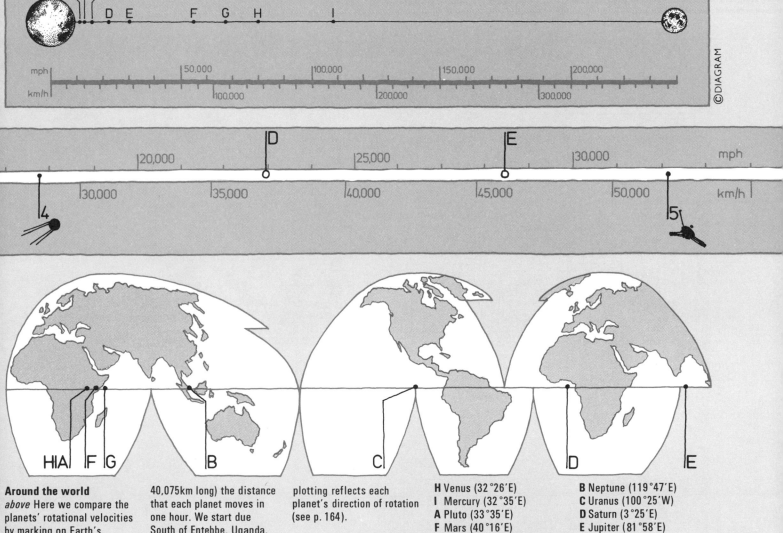

©DIAGRAM

Around the world
above Here we compare the planets' rotational velocities by marking on Earth's Equator (24,902mi/ 40,075km long) the distance that each planet moves in one hour. We start due South of Entebbe, Uganda, at longitude 32°29'E, and plotting reflects each planet's direction of rotation (see p. 164).

H Venus (32°26'E)
I Mercury (32°35'E)
A Pluto (33°35'E)
F Mars (40°16'E)
G Earth (47°31'E)

B Neptune (119°47'E)
C Uranus (100°25'W)
D Saturn (3°25'E)
E Jupiter (81°58'E)

The Sun has a rotational velocity at its equator of 4461mph. This is 1101 times faster than that of Venus, 4.3 times faster than that of Earth, and 6.3 times slower than that of Jupiter.

If Earth were to maintain its mean orbital velocity of 66,641mph but were to head toward the Sun, we would have only 58 days to make our escape!

CHAPTER 10

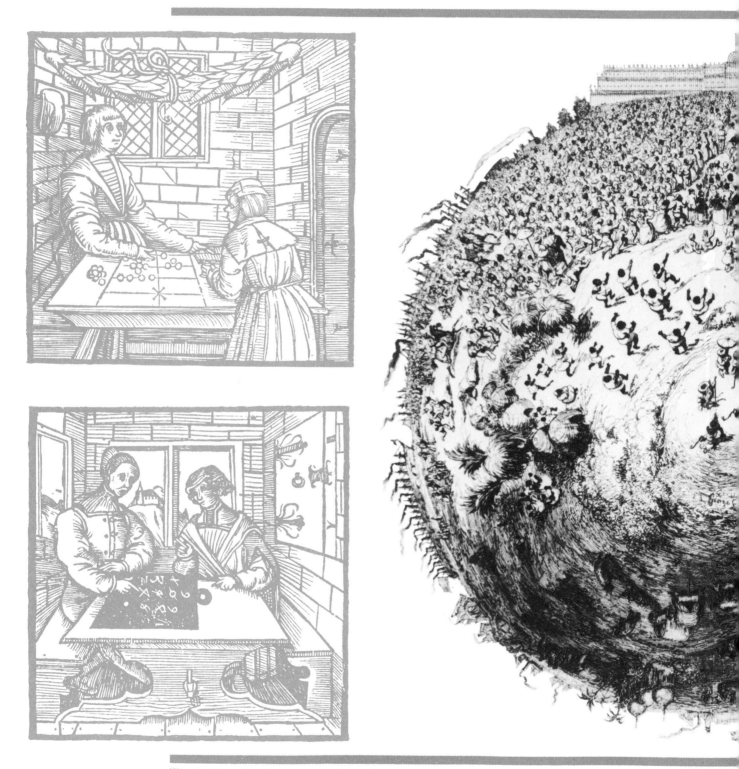

Woodcut illustrations from two books on arithmetic published in Augsburg in 1514. The first, from a work by Köbel, shows a simple line abacus, with ruled lines to show the different decimal orders and counters placed on or between the lines. The second, from a book by Böschensteyn, shows the working out of a problem with blackboard and chalk.

NUMBER

Vast numbers of people from all over the world are shown flocking to London's Great Exhibition of 1851 in George Cruikshank's frontispiece to *The World's Show* by Christopher Mayhew.

Lottery tickets from different countries make an interesting collage. The odds of winning a fortune are determined by the number of tickets sold.

NUMBERS

Here we look at the different number systems and number names that are basic to our perception of the world. Natural numbers, the first to have been used, are all the whole numbers from one to infinity. Integers consist of the natural numbers, zero, and also the minus numbers (−1, −2 etc.). Rational and real numbers include all the fractions and decimals between the whole numbers, while imaginary numbers are those that do not exist, such as the square root of a minus number. Numbers are the "tools" used for making and explaining all the other types of measurement devised by man and described in other chapters of this book.

International	US	Size	Indices
Ten	Ten	10	10^1
Hundred	Hundred	100	10^2
Thousand	Thousand	1000	10^3
Million	Million	1,000,000	10^6
Milliard	Billion	1,000,000,000	10^9
Billion	Trillion	1,000,000,000,000	10^{12}
	Quadrillion	1,000,000,000,000,000	10^{15}
Trillion	Quintillion	1,000,000,000,000,000,000	10^{18}

Couple	= 2
Pair	= 2
Half dozen	= 6
Dozen	= 12
Score	= 20
Quarter century	= 25
Half century	= 50
Century	= 100
Gross	= 144

Names and numbers *left* The table lists special names applied to some common amounts. Particular number systems are also used in specialized areas such as music (duet, trio, quartet etc.) and multiple births (twins, triplets, quadruplets etc.).

Counting in tens *above* In our numerical system we separate our numbers into groups by multiplying or dividing them by ten. This table gives the special names applied to some of these groups; there are two systems in current use—international and US.

Indicative indices *above* Indices are small numbers that indicate how many times one number must be multiplied by itself to give a second number. 100 (10 x 10) can be written as 10^2; 1000 (10 x 10 x 10) as 10^3; 1,000,000 as 10^6, and so on. Similarly 0.001 (which is $1 \div 10 \div 10 \div 10$) can be written as 10^{-3}, the minus sign indicating how many times the unit must be divided by ten.

Scale of natural numbers *above* Natural numbers are the whole numbers from one to infinity. This logarithmic scale represents the natural numbers that man uses to quantify the phenomena of his universe, some of which are listed by the scale (**K-R**) on the facing page.

The approximate number of molecules in one cubic centimeter of water is 33,000,000,000,000,000,000,000 (3.3 x 10^{22}).

An international billion is one thousand times bigger than a US billion. An international trillion is one million times bigger than a US trillion.

There are an estimated 10^{80} atoms in the universe.

Fixed quantities *right*
Although many quantities, such as the population of the world or the number of cars manufactured annually, are always changing, others remain constant. Here we list a few of these, which appear at the beginning of the scale *below*.

A	1 Empire State Building
B	2 sexes
C	3 sides to a triangle
D	4 playing card suits
E	5 fingers on a hand
F	6 legs on an insect
G	7 wonders of the world
H	8 legs on a spider
I	9 solar planets
J	10 Commandments

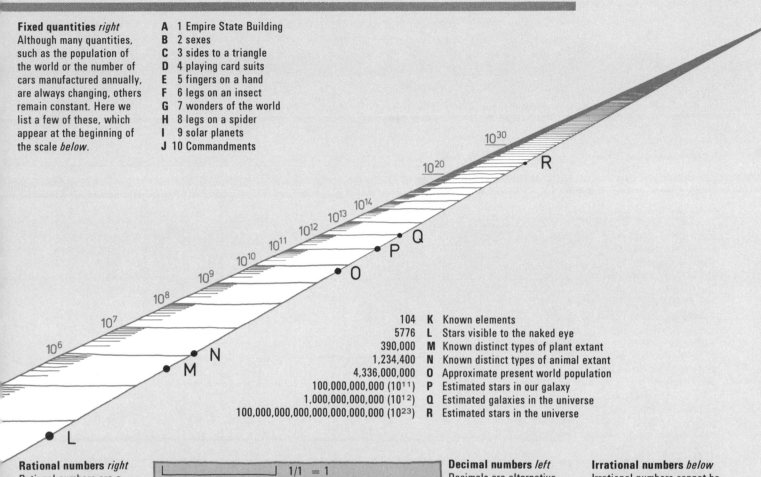

104	**K**	Known elements
5776	**L**	Stars visible to the naked eye
390,000	**M**	Known distinct types of plant extant
1,234,400	**N**	Known distinct types of animal extant
4,336,000,000	**O**	Approximate present world population
100,000,000,000 (10^{11})	**P**	Estimated stars in our galaxy
1,000,000,000,000 (10^{12})	**Q**	Estimated galaxies in the universe
100,000,000,000,000,000,000,000 (10^{23})	**R**	Estimated stars in the universe

Rational numbers *right*
Rational numbers are a means of expressing numbers as parts of a whole. The number below, or after, the dividing line is called the denominator, and tells us how many parts the whole has been divided into. The number above, or before, the line is called the numerator, and tells us how many of these parts we have. The principle is shown in the fractions listed in the first column of the table.

1/1	= 1
1/2	= 0.5
1/3	= 0.3333333333333333333
1/4	= 0.25
1/5	= 0.2
1/6	= 0.1666666666666666666
1/7	= 0.1428571428571428571
1/8	= 0.125
1/9	= 0.1111111111111111111
1/10	= 0.1
1/11	= 0.0909090909090909090
1/12	= 0.0833333333333333333
1/13	= 0.0769230769230769230
1/14	= 0.0714285714285714285
1/15	= 0.0666666666666666666
1/16	= 0.0625
1/17	= 0.0588235294110764705
1/18	= 0.0555555555555555555
1/19	= 0.0526315789473684210
1/20	= 0.05

Decimal numbers *left*
Decimals are alternative methods of expressing fractions that are not whole numbers. Each column to the right of the decimal point has one-tenth of the value of the previous column, so the number is divided into tenths, hundredths, thousandths etc. For example: 0.2 = 2/10; 0.03 = 3/100; 0.234 = 234/1000. The decimal equivalents of some fractions are shown in the table *left*. For some fractions, the decimal numbers stretch to infinity in a variety of repeating patterns.

Irrational numbers *below*
Irrational numbers cannot be expressed as fractions, and as decimals they reach to infinity without repeating the same series of numbers. Examples are π (the circumference of a circle ÷ twice the radius) and the square root of two.

$$\pi = 3.14159265$$

$$\sqrt{2} = 1.41421356$$

For one hundred thousand, the Indians have a quantity known as a *lakh*, written 1,00,000. Ten *lakhs* (equal to one million) is written 10,00,000. One hundred *lakhs* (equal to ten million) is termed a *crore*, and written 1,00,00,000.

NUMBER SYSTEMS

As civilizations developed, the exact quantification of items became necessary for purposes such as barter, tax, and calendar predictions of times and seasons. Many number systems were developed; the degree of sophistication varied, but each system had a base number that divided the units into manageable groups.

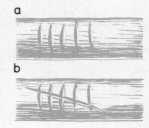

Making a mark *left*
The first method of making a record of an exact number of objects involved notching a stick or scratching a rock with one mark for each object (**a**). Sometimes groups of marks were indicated as shown (**b**).

A The Babylonians used a unit system, with only a few different symbols. Most numbers were made by building up series of these basic symbols in bases of ten and six, and by varying their placing.

B The Egyptians also used a unit system on a base of ten. Placing was variable; examples have been found written from left to right, right to left, and top to bottom.

C The Mayans used many systems, of which one is described here. It is a complex unit system based on five and twenty, in which placing is critical.

D The Hebrew system took letters of the alphabet and used these as symbols for their numbers. Although the system is on base ten, there are individual symbols for many of the higher numbers, such as 30 and 300.

E The Ionic system is also based on an alphabet. Like the Hebrew system, the Ionic is on base ten yet has symbols for many of the higher numbers.

F The Roman system is basically a unit system on bases of five and ten, but some of its symbols are the first letters of the names of of the numbers, for instance C for 100 (centum in Latin).

	0	1	2	3	4	5	6	7	8	9	10	50	100	500	1000
A															
B															
C															

	0	1	2	3	4	5	6	7	8	9	10	50	100	500	1000
D															
E		A	B	Γ	Δ	E	F	Z	H	Θ	I	N	P	Φ	/A
F		I	II	III	IV	V	VI	VII	VIII	IX	X	L	C	D	M

1980

Making a date!
Shown *left* is the date 1980 in modern Western numerals. The same date is depicted *right* in each of the number systems described *above*, keyed in by their identifying letters.

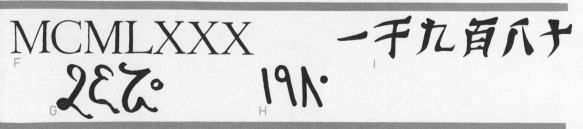

Here we represent 1980 with the dots and dashes of the International Morse Code used by telegraphists.

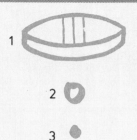

Zeroes *left*
Early number systems with a fixed value for every symbol had no need for zeroes. Zero symbols such as the Mayan (**1**), Hindu (**2**) and Arabic (**3**) permitted more compact number systems in which placing gives each numeral its value.

Use of the zero *right*
Zero allowed the indication of an empty column or line in a number, so keeping the other symbols in an exact value ratio. The examples show an addition with the use of zeroes (**a**), and the false result obtained when they are omitted (**b**).

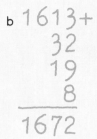

a
```
 10613+
  3200
   109
    80
 14002
```

b
```
 1613+
   32
   19
    8
 1672
```

G The Hindu system of numbering gives each number from zero to nine a symbol of its own with no other meaning—the symbols are not pictures or letters. Numbers above nine are built up on a base ten system using the basic symbols.
H The Arabic system, like the Hindu system, gives a symbol to every number from zero to nine, and builds up its higher numbers in the same way. This is the system on which Western numerals are based.
I Chinese numbering is a mixture of units (for the first three numbers) and individual symbols.

	0	1	2	3	4	5	6	7	8	9	10	50	100	500	1000
G	०	२	२	३	४	५	६	७	८	९	२०	५०	२००	५००	२०००
H	٠	١	٢	٣	٤	٥	٦	٧	٨	٩	١٠	٥٠	١٠٠	٥٠٠	١٠٠٠
I		一	二	三	四	五	六	七	八	九	十	五十	百	五百	千

$$1 \times 1000\ (10^3)\quad 9 \times 100\ (10^2)\quad 8 \times 10\quad 0 \times 1$$

1980

$$1 \times 1728\ (12^3)\quad 1 \times 144\ (12^2)\quad 9 \times 12\quad 0 \times 1$$

1190

$$1 \times 1024\ (2^{10})\quad 1 \times 512\ (2^9)\quad 1 \times 256\ (2^8)\quad 1 \times 128\ (2^7)\quad 0 \times 64\ (2^6)\quad 1 \times 32\ (2^5)\quad 1 \times 16\ (2^4)\quad 1 \times 8\ (2^3)\quad 1 \times 4\ (2^2)\quad 0 \times 2\quad 0 \times 1$$

11110111100

Base ten *above*
The base of a number system sets the points at which notation changes significantly in symbol or in placing. Our numbers are to base ten; a number in one column has ten times the value of the same number in the column to the right.

Base twelve *above*
If a number system is based on a factor of twelve, the units move into the left hand column at every twelfth unit instead of every tenth, and each column has the value of a power of twelve. The number 1980 is shown here to base twelve.

Binary *above*
The binary system uses a base of two. Each column has a value of a power of two, and the only numerals used are 0 and 1, since every two units require a move into the next column. Shown here is the number 1980 in binary notation.

MCMLXXX

F

G

H

一千九百八十

I

207

LINES AND SHAPES

Numbers can help us to define and reproduce specific shapes. All shapes are based on the concepts of points and the lines that join them. By counting the points and lines in a given shape and then reproducing exactly their lengths, contours and relationships to one another, duplicate shapes can be constructed.

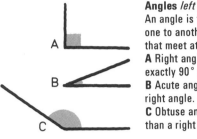

Angles *left*
An angle is the inclination one to another of two lines that meet at a point.
A Right angle, an angle of exactly 90° (see p. 210).
B Acute angle, less than a right angle.
C Obtuse angle, greater than a right angle.

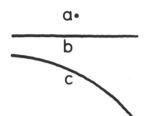

Lines and points *left*
Lines and points are in most people's minds inseparable from the marks used to represent them.
a Point, marked by a dot.
b Straight line, the shortest distance between two points.
c Curved line, a line that is nowhere straight.

Types of line *right*
Straight lines can be described in precise terms.
1 Horizontal, a level line.
2 Vertical, an upright line.
3 Oblique, a line inclining to the right or the left.
4 Parallel, two or more lines in the same direction that cannot meet.

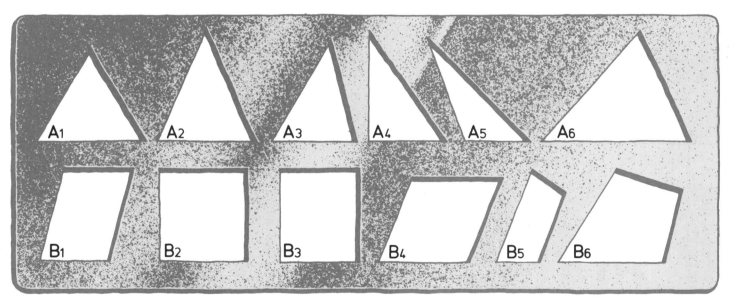

Triangles
A triangle is a figure contained by three straight sides. Some specific forms of triangle are illustrated *above* (**A**), and described in the list *far right*. The internal angles of any plane triangle always total exactly 180°.

Quadrilaterals
A quadrilateral is a figure contained by four straight sides; various types are shown *above* (**B**). If criteria listed in the table *right* are fulfilled, the shape will have have a particular name. Shapes outside these criteria are irregular.

A1 Equilateral triangle: all sides equal
A2 Isosceles triangle: two equal sides
A3 Scalene triangle: no two sides equal
A4 Right-angled triangle: contains one right angle
A5 Obtuse-angled triangle: contains one obtuse angle
A6 Acute-angled triangle: contains three acute angles
B1 Parallelogram: opposite sides are parallel
B2 Square: equal sides and four right angles
B3 Rectangle: equal opposite sides and four right angles
B4 Rhombus: equal sides but no right angles
B5 Trapezium: two parallel sides
B6 Irregular: no parallel sides

A triangle drawn on a sphere breaks the Euclidian law that the internal angles of a triangle equal 180°. Our example has three right angles (270°).

The ancient Egyptians obtained a right angle with a piece of rope divided into 12 equal sections by 12 knots. One man (**a**) held the first knot and the other end of the rope. A second man (**b**) held the knot three sections from the first one. A third man (**c**) held the knot five spaces after the second man's knot and four from the end.

Circle *left*
A circle is a plane figure contained by one curved line, every point of which is equally distant from the point at the circle's center. The line bounding the circle is called the lcircumference.

Features of a circle *right*
a Radius, a straight line joining the center and any point on the circumference.
b Diameter, a straight line joining opposite points on the circumference.
c Sector, the area between any two radii.
d Arc, the line along the circumference joining any two points.
e Chord, the straight line joining any two points on the circumference.
f Segment, the area bounded by an arc and a chord.
g Tangent, a straight line touching the circumference at right angles to a radius.

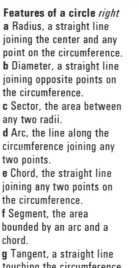

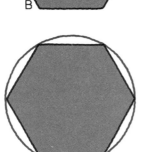

Polygons *above*
A polygon is a plane figure bounded by straight lines. If its sides or angles are unequal, the polygon is irregular (**A**); if they are equal, it is regular (**B**) and will fit into a circle with all its vertices touching the circumference (*right*).

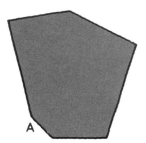

Regular polygons
Listed in the table *below* are regular polygons, from three-sided to twelve-sided. The table gives their correct names, the number of sides that each one possesses, the size of the internal angles made by the sides where they join around the perimeter, and the sum of these angles. The diagram *right* shows the same polygons with sides of equal length throughout. As the number of sides increases, so do the size of the individual angles and the total of the internal angles.

Name of polygon	Sides	Each angle	Sum of angles
Triangle	3	60°	180°
Square	4	90°	360°
Pentagon	5	108°	540°
Hexagon	6	120°	720°
Heptagon	7	128.6°	900°
Octagon	8	135°	1080°
Nonagon	9	140°	1260°
Decagon	10	144°	1440°
Undecagon	11	147.3°	1620°
Dodecagon	12	150°	1800°

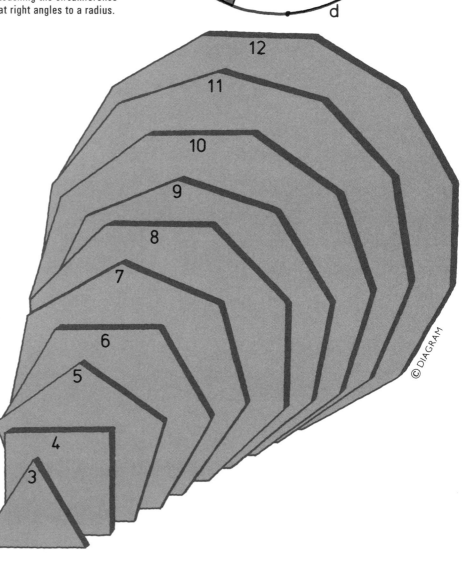

© DIAGRAM

ANGLES

On this page we compare three systems of measuring angles, all based on divisions of a circle. The original system of degrees was devised by the ancient Mesopotamians, and is still the most widely used system today. The radian is favored by modern mathematics; the grade, or gon, developed in 1792, is a metric measurement of angle.

Degrees in a circle *right*
Here we give the names of some regular divisions of the circle, based on the degree system. The degree is subdivided into the minute, and the minute into the second, in the ratios indicated.

360°	= 1 circle
90°	= 1 right angle
60°	= 1 sextant
45°	= 1 octant
30°	= 1 sign
1°	= 60 minutes (60')
1'	= 60 seconds (60")
21,600'	= 1 circle
1,296,000"	= 1 circle

Degrees *right, below*
The degree system divides the circle into 360 degrees (360°). If XY is a quarter of a circle, its angle XZY in degrees is 360 divided by four, which equals 90° (known as a right angle). A semicircle has an angle of 180°.

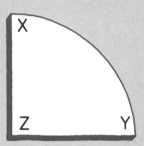

Radians *right, below*
A radian is the angle at the center of a circle that cuts off an arc on the circumference equal in length to the radius. If the arc XY equals the radii ZX and ZY, the angle XZY equals one radian (1 rad), and 1 rad equals 57.2958°.

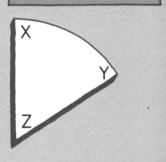

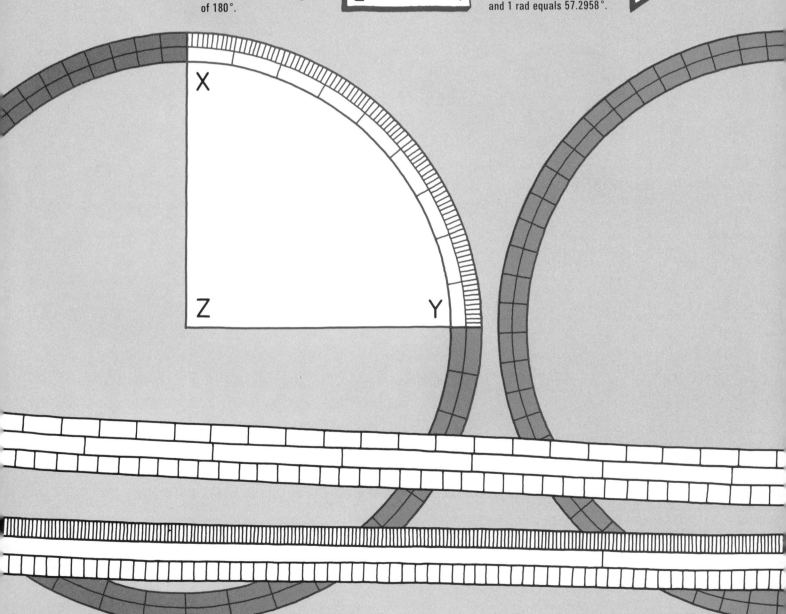

A good target rifle can place all its shots to within one minute of angle. This means that at 1000 yards it is able to hit a 10-inch diameter bullseye with every shot.

360° = 6.283183 rads
1 rad = 100 centirads
1 rad = 1000 millirads
1 rad = 1000 angular mils
360° = 400 grades
1 grade = 100 centigrades
1 grade = 100 new minutes
1 grade = 1000 milligrades
1 grade = 10,000 new seconds

Radians and grades *left*
The list gives the number of radians and grades in a circle, and also the names and values of some of their most common subdivisions.

Grades *right, below right*
A grade, or gon, is one-hundredth of a right angle. When XY is a quarter of the circumference, then the angle XZY is equal to 100 grades. Each circle contains 400 grades, and one grade is equal to 0.9°.

Degrees		Centirads	Degrees		Grades
0.5730	1	1.7453	0.9	1	1.1111
1.1459	2	3.4907	1.8	2	2.2222
1.7189	3	5.2360	2.7	3	3.3333
2.2918	4	6.9813	3.6	4	4.4444
2.8648	5	8.7266	4.5	5	5.5555
3.4378	6	10.472	5.4	6	6.6667
4.0107	7	12.217	6.3	7	7.7778
4.5837	8	13.963	7.2	8	8.8889
5.1566	9	15.708	8.1	9	10.000
5.7296	10	17.453	9.0	10	11.111
57.296	100	174.53	90.0	100	111.11
114.59	200	349.07	180.0	200	222.22
171.89	300	523.60	270.0	300	333.33
			360.0	400	444.44

Conversion tables *left*
These tables are for converting degrees to centirads and grades, and vice versa. To use them, first find the figure to be converted in the central column of the relevant table; its equivalent can then be found in the appropriate column to the right or the left as indicated by the headings above.

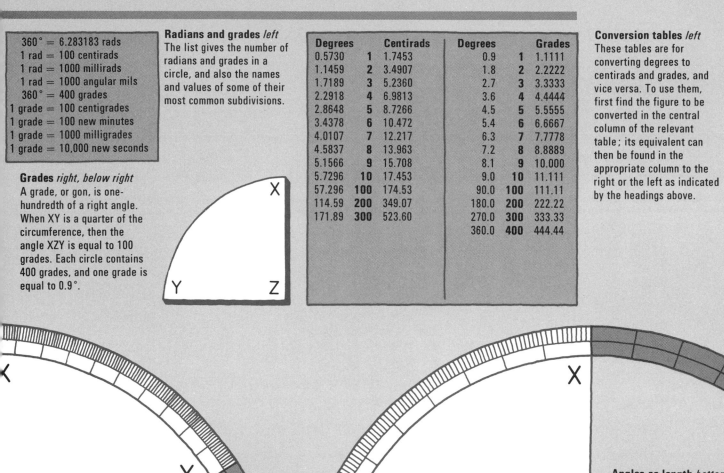

Angles as length *bottom left* These two sets of three lines represent parts of the circumference of two circles; the first (**ABC**) of radius 112.9ft (34.4m) and the second (**DEF**) of radius 676.8ft (206.3m). On these circles are marked the lengths of:
A Minutes (1′ = 10mm)
B Centirads (1c = 34.4mm)
C Centigrades (1cg = 5.4mm)
D Seconds (1″ = 1mm)
E Millirads (1m = 206.24mm)
F Milligrades (1mg = 3.24mm)

SOLID FORMS

Three-dimensional solid forms have area, volume and depth. It is the addition of depth that distinguishes them from two-dimensional geometric shapes. Any solid form made up of plane faces is given the general name of "polyhedron." Solids that include curved surfaces are named according to the shape of the curve.

Non-regular solids
left Some of the most common are shown here.
1 Sphere
2 Spheroid
3 Torus
4 Cone
5 Cylinder
6 Prism
7 Pyramid (if its base has four or more sides)

Containing a regular solid
below One property of a regular solid is that when it is placed in a sphere, all its corners will touch the sphere.

Regular solids *right*
If the surfaces of a solid are regular polygons of equal size, it is called a regular solid. There are only five possible regular solids and of these, three have surfaces that are equilateral triangles. Here we illustrate the regular solids and also show you how they can be made by cutting and folding a piece of card. Names of the regular solids are given here with the number of sides in brackets.
A Tetrahedron (4)
B Hexahedron or cube (6)
C Octahedron (8)
D Dodecahedron (12)
E Icosahedron (20)

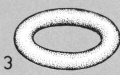
1

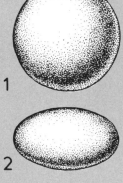
2

3

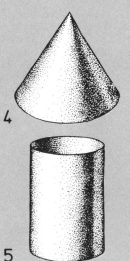
4

5

6

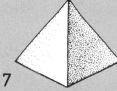
7

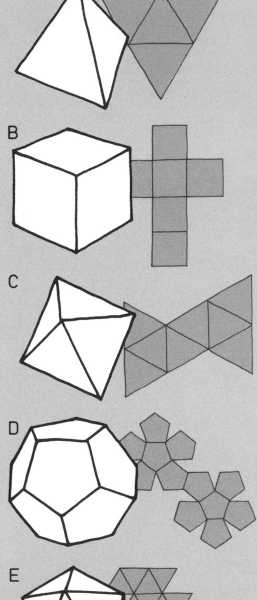

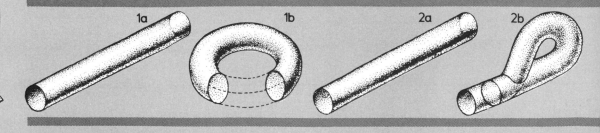
1a   1b   2a   2b

Followers of Pythagoras in Ancient Greece believed that earth is made of cubes, air of octahedrons, fire of tetrahedrons and water of icosahedrons. The dodecahedron was the symbol of the universe as a whole.

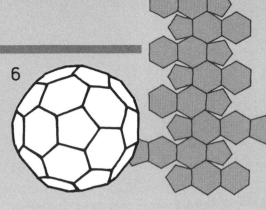

Semi-regular solids

right Semi-regular solids have surfaces or faces composed of more than one type of polygon. For example, the cuboctahedron (**2**) has faces that are squares and equilateral triangles. The truncated icosidodecahedron (**8**) has squares, hexagons and decagons. Like regular solids, all semi-regular solids fit into a sphere.

For any solid form with flat faces, the number of edges equals the number of faces plus the number of corners minus two. For example the cuboctahedron has 12 corners and 14 faces; it therefore has 24 edges.

A selection of semi-regular solids is included here.

1 Truncated octahedron or mecon
  24 corners, 14 faces
2 Cuboctahedron or dymaxion
  12 corners, 14 faces
3 Truncated cuboctahedron
  48 corners, 26 faces
4 Snub cube
  24 corners, 38 faces
5 Rhombicuboctahedron or square spin
  24 corners, 26 faces
6 Truncated icosahedron
  60 corners, 32 faces
7 Icosidodecahedron
  30 corners, 32 faces
8 Truncated icosidodecahedron
  120 corners, 62 faces
9 Snub dodecahedron
  60 corners, 92 faces
10 Rhombicosidodecahedron
  60 corners, 62 faces

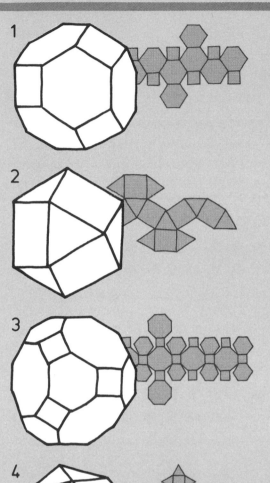

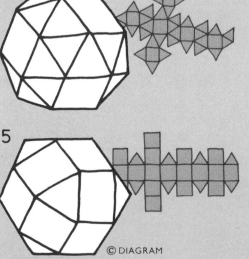

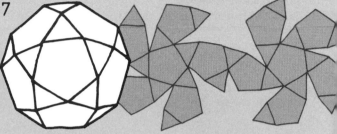

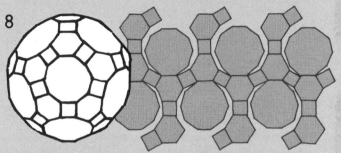

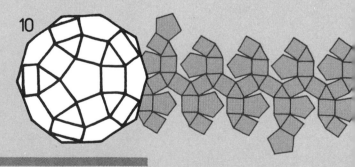

© DIAGRAM

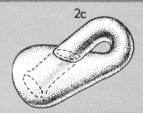

2c

When inside is outside

left Compared are two forms made by joining the ends of a cylinder.
1 If we join the ends of a cylinder (**a**) in the usual way, we form a torus (**b**), because the ends point in opposite directions when they meet.
2 For a Klein bottle both ends of the cylinder (**a**) must be pointing in the same direction when they meet. It is made by putting one end through the wall of the cylinder (**b**), and then joining it to the other end from the outside (**c**). This extraordinary solid has only a single surface, and thus no "inside" or "outside."

NUMBER SHAPES

If every whole number is represented by the equivalent number of dots, the numbers fall into groups or "series" in which each number can be formed into a certain geometric shape, such as a square or a pyramid. The general term for these series is "polygonal numbers." Some of the simple series are shown on these two pages.

Comparing the series *right*
The table shows the series to which each number from one to twenty belongs. One belongs to five different series, four and ten to four series, and the other numbers to either three or two series.

Even
Odd
Rectangular
Prime
Square
Triangular
Pyramidal
Tetrahedral

Even numbers *left*
All numbers that can be divided by two are called "even" numbers. We can show this series of numbers geometrically by beginning with a block of two, and adding a similar block of two for every successive even number.

Odd numbers *left*
All those numbers that cannot be divided by two without a single unit remaining are called "odd" numbers. This series is built by beginning with a single dot and then adding a block of two for every successive odd number.

Rectangular numbers *left*
Some numbers can be divided into two or more equal parts that can be arranged to form a rectangle. Rectangular numbers are always the product of (the result of multiplying) two smaller numbers, for instance 8 = 4 x 2.

Prime numbers *left*
A prime number is a number greater than one that cannot be divided by any number other than one and itself. This means that prime numbers cannot be fitted into square or rectangular shapes.

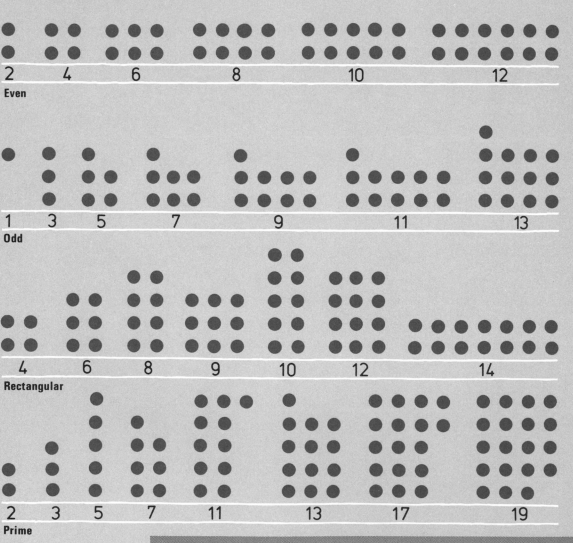

Even: 2 4 6 8 10 12

Odd: 1 3 5 7 9 11 13

Rectangular: 4 6 8 9 10 12 14

Prime: 2 3 5 7 11 13 17 19

A "perfect" number is a number of which the sum of all its divisors exactly equals itself. An example is 6 (of which the divisors are 1, 2 and 3). There are only seven perfect numbers between 1 and 40,000,000 (6; 28; 496; 8128; 130,816; 2,096,128; 33,550,336).

From 1 to 20 there are 11 rectangular numbers, 10 odd numbers, 10 even numbers, 8 prime numbers, 5 triangular numbers, 4 square numbers, 4 tetrahedral numbers, and 3 pyramidal numbers.

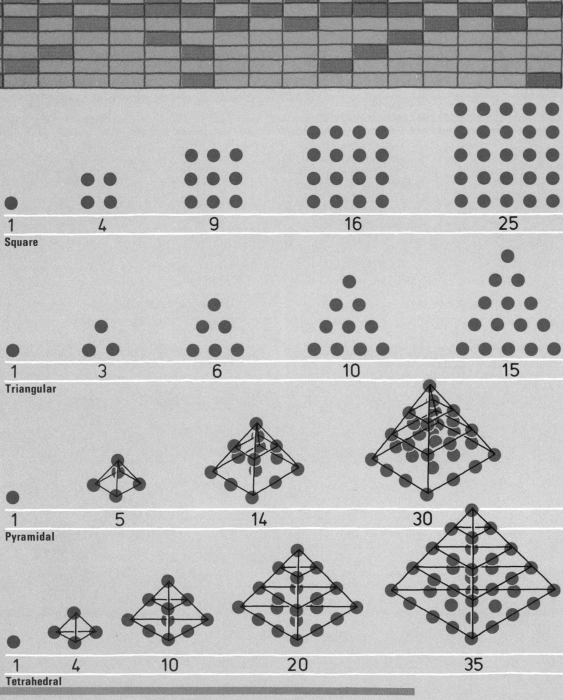

Square numbers *right*
When we multiply a number by itself, we say that we have squared it. This is because the product of a number multiplied by itself can always be formed into a square, each side of which will represent the original number in dots.

Triangular numbers *right*
Triangular numbers can be formed into regular triangles, with the same number of dots on each side. To find the second number, two is added to the previous number (1+2 = 3); to find the third, three is added to the second, and so on.

Pyramidal numbers *right*
These numbers can be formed into pyramids with five surfaces. Each pyramidal number is made up of layers of square numbers. 1 is first, followed by 1+4, then 1+4+9; this sequence continues throughout the series.

Tetrahedral numbers *right*
Tetrahedrons are shapes with four surfaces, and tetrahedral numbers fit into these shapes. As shown in the illustration, they are made up of layers of triangular numbers; first 1, then 1+3, then 1+3+6, then 1+3+6+10, and so on.

Square
1   4   9   16   25

Triangular
1   3   6   10   15

Pyramidal
1   5   14   30

Tetrahedral
1   4   10   20   35

Cannonballs were commonly stacked in pyramids. A stack 10 high would contain a total of 385.

©DIAGRAM

ANIMAL LIMBS

Many creatures have legs that serve as props and as levers aiding locomotion. Nature produces such limbs in an astonishing variety of numbers and designs. Some limbs that support or move a body are not technically speaking legs, but rather arms or feet. But we include some here because of their locomotive function.

Legs roll call
The beasts *right* and *below* are ranged in order of number of legs (indicated for each animal). Most of those with six or more legs are small members of that great backboneless group, arthropods, creatures whose name means jointed legged.

1 Edible snail, a mollusk that crawls on one muscular foot. Its muscles thrust back on the ground to push the snail forward.
2 Flamingo, a biped (two-legged like all other birds and man). Long, stilt-like legs help it to wade in shallow lagoons.

3 Tripod fish. Two long rays from the pectoral fins and one from the tail stop this deep-sea fish sinking when it rests on soft mud.
4 Giraffes and most other mammals, reptiles, and amphibians are tetrapods: animals stably supported on four legs and feet.

5 Most starfish have five limbs called arms. Starfish haul themselves along on many tiny tube feet poking out beneath each arm.
6 The human louse is an insect. This huge group of six-legged arthropods includes ants, beetles, bees, flies and moths.

7 Springtails are insects, and so have six true legs. But a forked springing organ at the hind end acts as a kind of seventh leg.
8 Harvestmen are arachnids, eight-legged arthropods related to the scorpions and spiders. (The 8-armed octopus is a mollusk.)

9 A 9-armed sunstar. This type of starfish can have up to 50 arms.
10 Shrimp. A decapod (10-footed) crustacean, as are crabs and lobsters.
12 Sunstar with 12 arms. Individuals of the species *Solaster papposus* can have 8–13 arms.

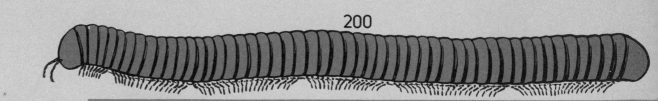

200

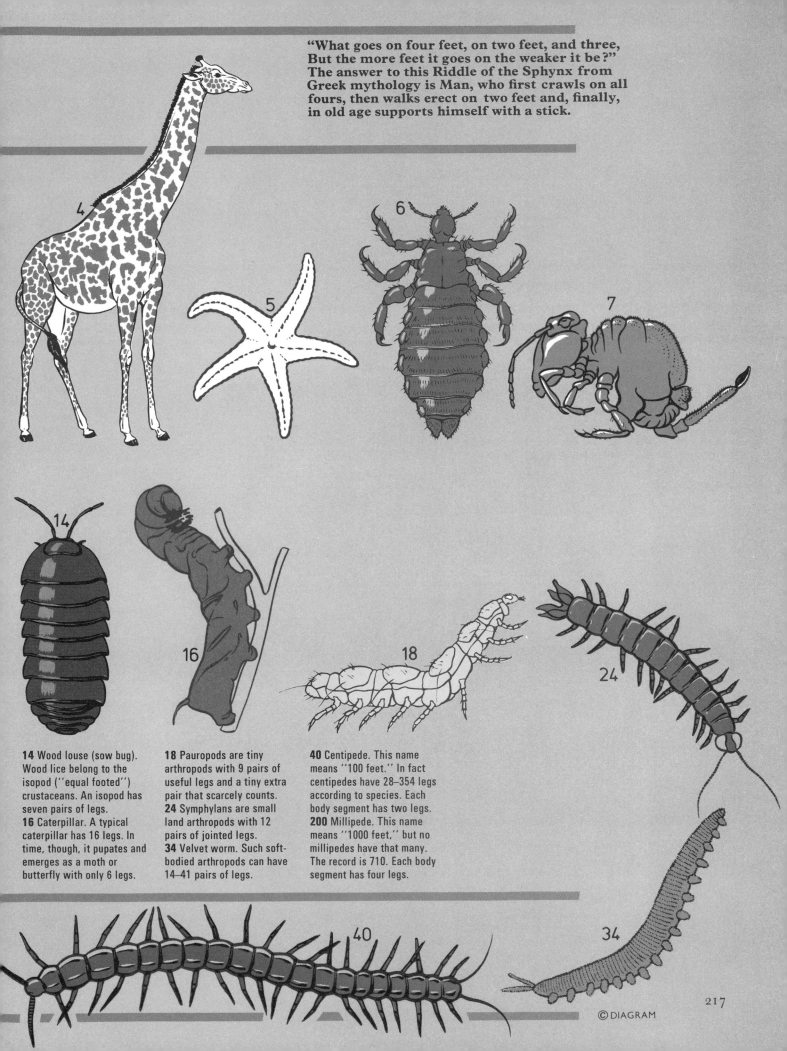

"What goes on four feet, on two feet, and three, But the more feet it goes on the weaker it be?" The answer to this Riddle of the Sphynx from Greek mythology is Man, who first crawls on all fours, then walks erect on two feet and, finally, in old age supports himself with a stick.

14 Wood louse (sow bug). Wood lice belong to the isopod ("equal footed") crustaceans. An isopod has seven pairs of legs.
16 Caterpillar. A typical caterpillar has 16 legs. In time, though, it pupates and emerges as a moth or butterfly with only 6 legs.

18 Pauropods are tiny arthropods with 9 pairs of useful legs and a tiny extra pair that scarcely counts.
24 Symphylans are small land arthropods with 12 pairs of jointed legs.
34 Velvet worm. Such soft-bodied arthropods can have 14–41 pairs of legs.

40 Centipede. This name means "100 feet." In fact centipedes have 28–354 legs according to species. Each body segment has two legs.
200 Millipede. This name means "1000 feet," but no millipedes have that many. The record is 710. Each body segment has four legs.

217

©DIAGRAM

ANIMAL OFFSPRING

Each kind of creature tends to produce enough eggs or young to keep up the numbers of its species. The elephant—a long-lived mammal with few enemies—bears only one calf in about two years. At the opposite extreme is the evidently vulnerable ocean sunfish, which lays 300 million eggs in a single spawning.

Mammals

Births per mother commonly range from one in elephants (**A**) to 13 in opossums (**B**). Tenrecs, pigs and mice have all produced record litters of over 30. The human record for multiple births is 10—but no babies survived.

Birds

Clutch size ranges from one in the emperor penguin (**C**), to 15 in the ostrich (**D**). One limiting factor is the surplus food needed by each egg-layer. In two weeks a blue tit eats enough to lay a 10-egg clutch that weighs more than she does.

Reptiles

Most reptiles lay eggs but some give birth to living young. Eggs or babies number from one to several dozen. Few lizards can rival the common agama (**E**), laying an average of 16.6 and up to 23 eggs. Even more prolific are some snakes. Pythons (**F**) average a clutch of 29 but may exceed 100. The reptiles' record is held by marine turtles. When a female green turtle (**G**) hauls ashore, on average she lays 104 eggs; the listed peak is 184. Nile crocodiles (**H**) lay some 60 eggs and a Mississippi alligator has laid 88.

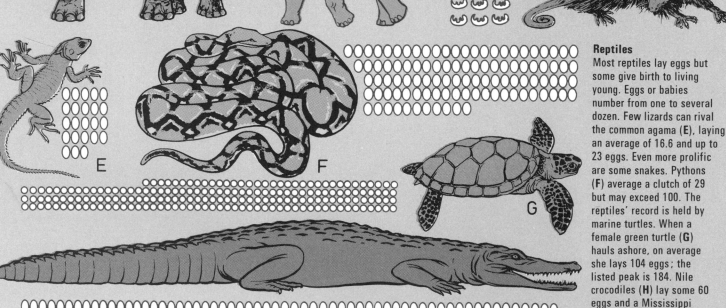

In her life a green turtle lays an average of 1800 eggs. Of these, some 1395 don't hatch, 374 hatchlings quickly die, and only 3 live long enough to breed.

The highest officially recorded number of children born to one mother is 69, to a Russian peasant woman in the 18th century. They comprised four sets of quadruplets, seven sets of triplets, and 16 pairs of twins.

In favorable conditions *Escherichia coli*, a bacterium of the human gut, splits in two every 15 minutes to reproduce. In 24 hours one bacterium can become 4×10^{28}.

Number

Amphibians

Frogs and toads lay eggs in thousands. For the giant toad (**I**), 35,000 is a normal maximum. Tailed amphibians produce smaller totals. The hellbender (**J**) leads the field, with a chain of up to 450 eggs like loosely strung pearls.

Fishes

In one spawning an ocean sunfish (**K**) may lay 300 million eggs, so 15 sunfish could lay as many eggs as there are people. Yet some kinds of ray (**L**) lay as few as two, big, yolky eggs. Tough outer cases help the eggs to survive.

Insects

A queen termite (**M**) can lay 8000 eggs per day for years, thanks to an abdomen that grows into a swollen egg-laying machine. But a potter wasp (**N**) laboriously shapes a hollow ball of mud in which she lays only one egg.

Crustaceans and mollusks

As many as 1,750,000 eggs per ''clutch'' make the blue crab (**O**) among the most fertile of crustaceans. But some mollusks are even more prolific. An edible mussel (**P**) squirts up to 25 million eggs into the sea in a single spawning.

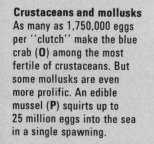

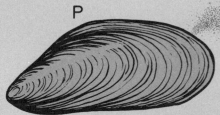

219

©DIAGRAM

POPULATION 1

Of the people alive today approximately 64% live in Asia, 11.1% in Europe, 10.5% in Africa, 8.4% in North America, 5.5% in South America and 0.5% in Oceania. On these two pages we look at the historical growth of population, at population by continent, and at the largest national populations in the world and in Europe.

World population today
Listed *right* and used for the diagram *below* are the most recent UN world and continental population figures available (mid-1979 provisional estimates). Asia including the USSR has 64% of the world total, without the USSR it has 58%.

World	4,336,000,000
a Asia*	2,773,000,000
b Europe	482,000,000
c Africa	456,000,000
d N America	364,000,000
e S America	239,000,000
f Oceania	22,000,000

*Includes USSR

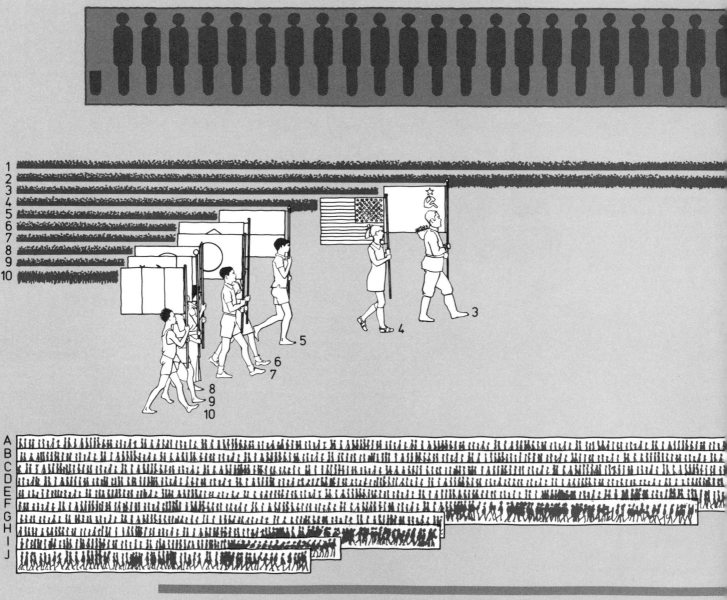

If the entire population of China were acrobats standing one on another's shoulders (each adding 4ft to the height of the column) they would extend out into space more than three times farther than the distance from Earth to Moon.

The estimated population of the world increased by 2.7 times in the 250 years from 1650 to 1900. In less than one-third of this time, from 1900 to the present day, it has increased by a further 2.9 times.

World population growth
right Shown here are world population estimates for dates from 1650 to the present. The diagram's sweeping curve is due to the cumulative nature of population increase and, especially this century, to medical advances.

1650	550,000,000
1700	600,000,000
1750	725,000,000
1800	900,000,000
1850	1,200,000,000
1900	1,500,000,000
1979	4,336,000,000

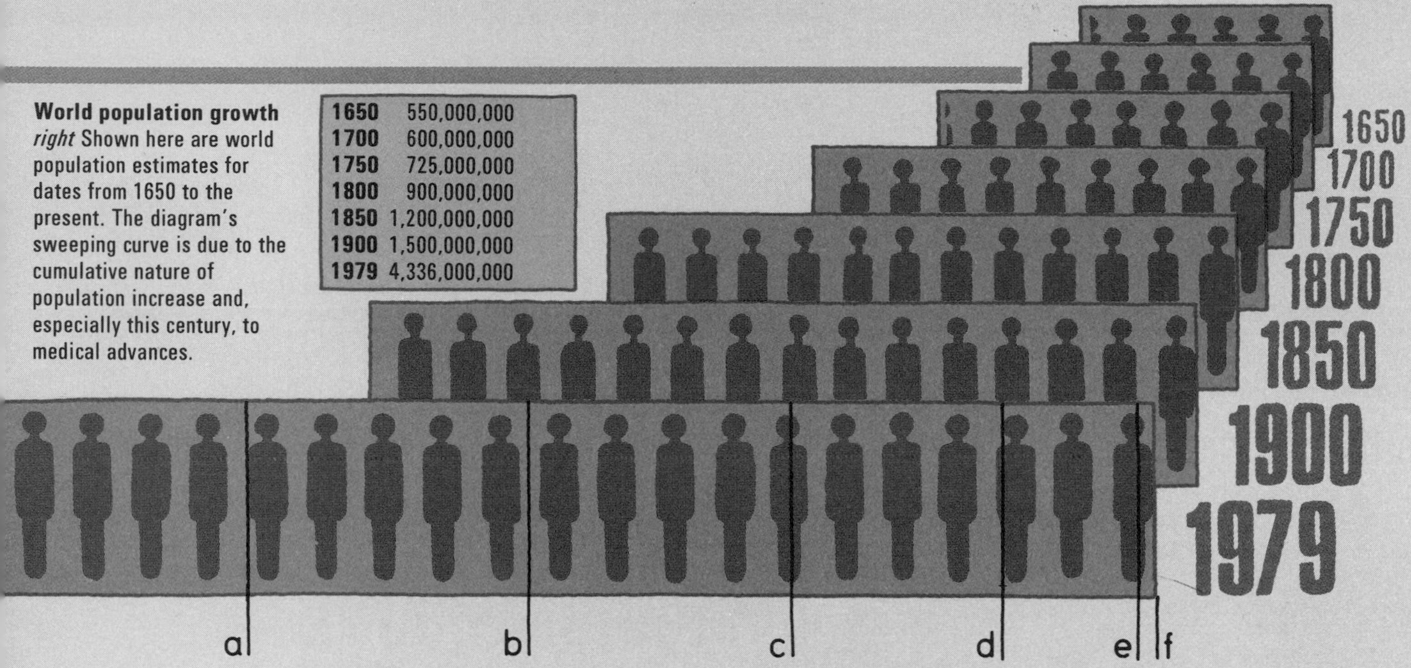

1650
1700
1750
1800
1850
1900
1979

a b c d e f

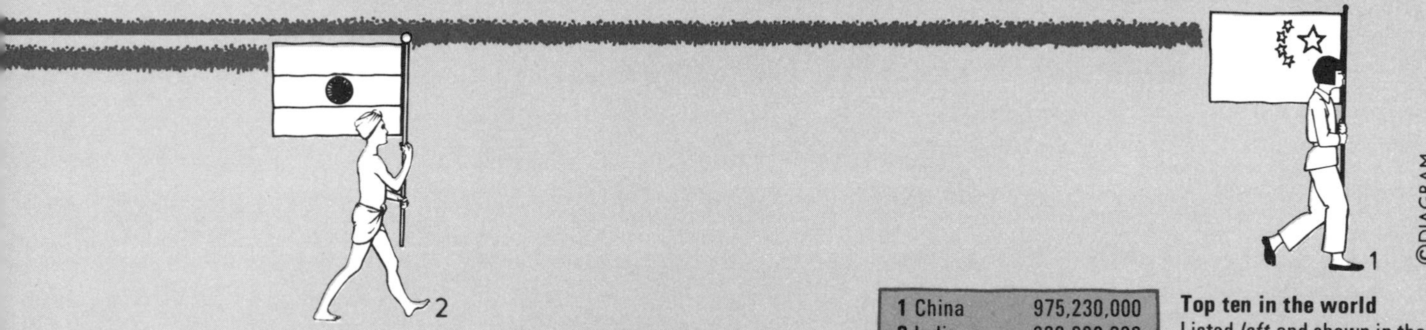

©DIAGRAM

1	China	975,230,000
2	India	638,388,000
3	USSR	261,569,000
4	USA	218,059,000
5	Indonesia	145,100,000
6	Brazil	115,397,000
7	Japan	114,898,000
8	Bangladesh	84,655,000
9	Pakistan	76,770,000
10	Nigeria (1978)	72,220,000

Top ten in the world
Listed *left* and shown in the diagram *above* are the countries with the 10 largest populations (UN estimates, published 1979). China, which ranks first in terms of population, has half as many people again as India, which ranks second.

A	West Germany	61,340,000
B	Italy	56,697,000
C	Great Britain	55,822,000
D	France	53,324,000
E	Spain	37,109,000
F	Poland	35,133,000
G	Yugoslavia	22,083,000
H	Romania	21,855,000
I	East Germany	16,760,000
J	Czechoslovakia	15,140,000

Top ten in Europe *left*
Listed in the table and shown in the diagram are the 10 European countries with the largest populations (UN estimates, published 1979–80). The top four combined exceed the population of the USA by more than nine million.

During the 1970s world population increased annually at a rate of 1.9%. If this same rate is maintained until the end of the century, world population in the year 2000 will be approximately 6400 million, an increase of 48% compared to the present day.

POPULATION 2

On these two pages we look at population figures for some of the world's largest cities. Different sources publish widely differing figures for city populations—a consequence of the problems involved in differentiating the population of a city itself from that of its suburbs. We use the most recent UN figures (mostly for the mid-1970s).

1	Mexico City	8,628,000
2	Tokyo	8,592,000
3	New York	7,482,000
4	Sao Paulo	7,199,000
5	London	7,028,000
6	Moscow	6,942,000
7	Seoul	6,879,000
8	Bombay	5,971,000
9	Cairo	5,921,000
10	Rio de Janeiro	4,858,000

World's largest cities
right Listed in the table and located on the map are the world's 10 most populous cities. Symbols of people (1 = 200,000) show their comparative populations. The largest, Mexico City, has twice as many people as Norway.

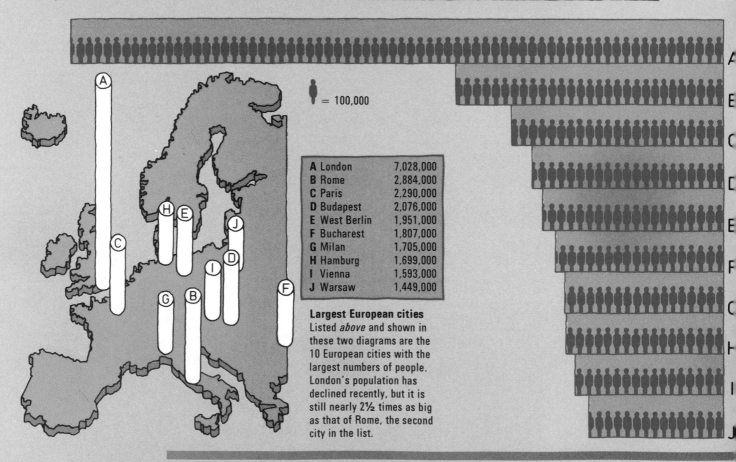

= 100,000

A	London	7,028,000
B	Rome	2,884,000
C	Paris	2,290,000
D	Budapest	2,076,000
E	West Berlin	1,951,000
F	Bucharest	1,807,000
G	Milan	1,705,000
H	Hamburg	1,699,000
I	Vienna	1,593,000
J	Warsaw	1,449,000

Largest European cities
Listed *above* and shown in these two diagrams are the 10 European cities with the largest numbers of people. London's population has declined recently, but it is still nearly 2½ times as big as that of Rome, the second city in the list.

Of London's estimated 7,028,000 inhabitants, only about 5600 (0.08%) live in the municipal area officially known as the "City of London."

If all the inhabitants of Mexico City were to stand side by side (allowing 18 inches per person) they would cover a distance of 2451 miles, only 3 miles short of the Great Circle distance from Mexico City to Quebec.

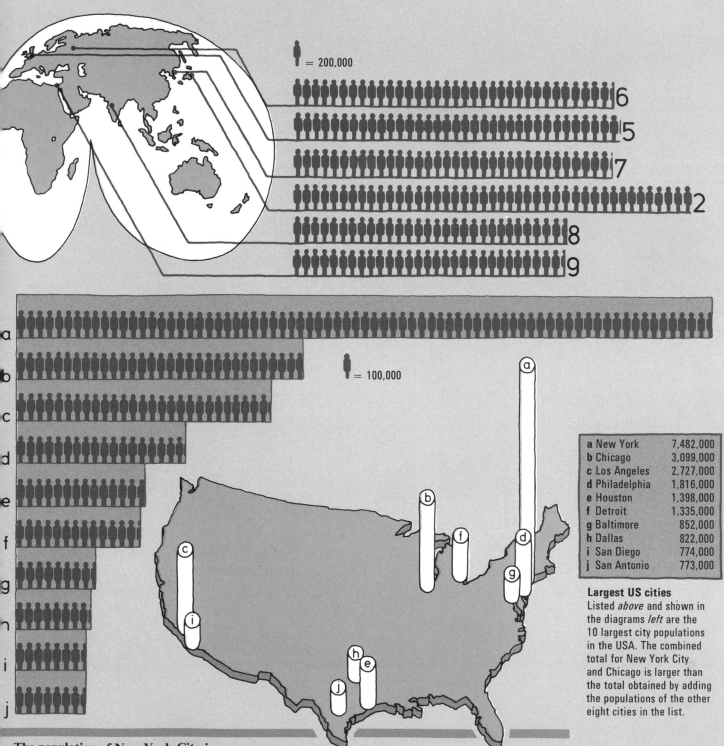

♂ = 200,000

♂ = 100,000

a	New York	7,482,000
b	Chicago	3,099,000
c	Los Angeles	2,727,000
d	Philadelphia	1,816,000
e	Houston	1,398,000
f	Detroit	1,335,000
g	Baltimore	852,000
h	Dallas	822,000
i	San Diego	774,000
j	San Antonio	773,000

Largest US cities
Listed *above* and shown in the diagrams *left* are the 10 largest city populations in the USA. The combined total for New York City and Chicago is larger than the total obtained by adding the populations of the other eight cities in the list.

The population of New York City is approximately 41% of that of New York State, and approximately 3% of that of the entire USA.

© DIAGRAM

SPORTS TEAMS

Man has devised a great variety of sporting activities for different numbers of participants. Here we concentrate on the numbers of players involved in different team sports—from the maximum of 43 named players for a professional game of American football to the four riders, excluding substitutes, required for polo.

Playing with numbers *right*
A professional American football team (1) can have up to 43 players in uniform, but only 11 of them are allowed on the field at any time. An ice hockey team (2) consists of 17 players, of whom only six may be on the ice at the same time.

Sport	A	B	C
a Australian rules football	18	2	
b Gaelic football	15		3
c Hurling	15		3
d Rugby union	15		2
e Rugby league	13	2	
f Canadian football	12	*	
g Lacrosse (women's)	12		*
h Korfball	12	*	
i American football	11	32	
j Speedball	11	5	
k Bandy	11	3	
l Soccer	11	2	
m Cricket	11		*
n Field hockey	11	2	
o Lacrosse (men's)	10	9	
p Baseball	9	*	
q Softball	9	6	
r Rounders	9		*
s Team handball	7	5	
t Water polo	7	4	
u Netball	7		*
v Ice hockey	6	11	
w Volleyball	6	6	
x Basketball	5	7	
y Roller hockey	5	5	
z Polo	4		*
A Maximum number per team allowed on the field			
B Substitutes permitted for any reason			
C Substitutes to replace sick or injured players			
* Actual number unspecified in official rules			

Players and substitutes
The first column of figures in the table *above* and the large numbers beneath the illustrations *right* indicate the maximum numbers of players in different types of team who are allowed on the field, or other playing area, at any one time in a game. The other two columns in the table show maximum numbers of permitted substitutes, differentiating between their use for tactical purposes and their more limited use merely to replace players who are incapacitated during play.

If a team of men wanting to play lacrosse to international rules had mistakenly bought the rule book for the women's game, they would find themselves with two too many players on the field, but an insufficient number of substitutes.

The 43 named players for a professional American football game could be divided instead into teams (without substitutes) for all of the following: soccer, cricket, baseball, volleyball and ice hockey.

© DIAGRAM

n	o	p	s	v	x	z

11 10 9 7 6 5 4

i,n	o	p	s	v	x	z

The largest number of spectators ever to crowd into a stadium to watch any game was 199,854, for the Brazil versus Uruguay soccer international in Rio de Janeiro on July 16, 1950. This total works out at some 9084 spectators for every player on the field.

VEHICLE CAPACITIES

In developing the best vehicle to meet a particular transportation requirement, designers must strike a balance between the numbers of people likely to make use of the service, and other factors such as safety, comfort and speed. Here we compare the numbers of people carried by various types of land, air and water vehicle.

Land transport *right*
Here we show a selection of land vehicles together with the number of passengers they are authorized to carry.
a London taxi, 4
b Greyhound bus, 43
c London bus (RML), 72
d World's longest bus, US-built for Middle East, 187

Air transport *right*
Combined passenger and crew capacities are here shown for a range of civil aircraft, from executive jet to jumbo.
1 Beechcraft H18, 11
2 *Concorde*, 148
3 Boeing 707, 221
4 Boeing 747, 500

Water transport *right*
Drawn here is a selection of vessels, with their combined complements of passengers and crew.
A Venetian gondola, 5
B *Mayflower*, about 130
C Commercial hovercraft, SR-N4/Mk 3, 428
D *Queen Elizabeth 2*, 2931

A Venetian with a gondola able to take four passengers at a time would have to make 733 journeys to ferry out the crew and passengers of the QE2.

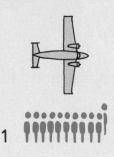

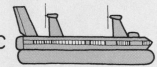

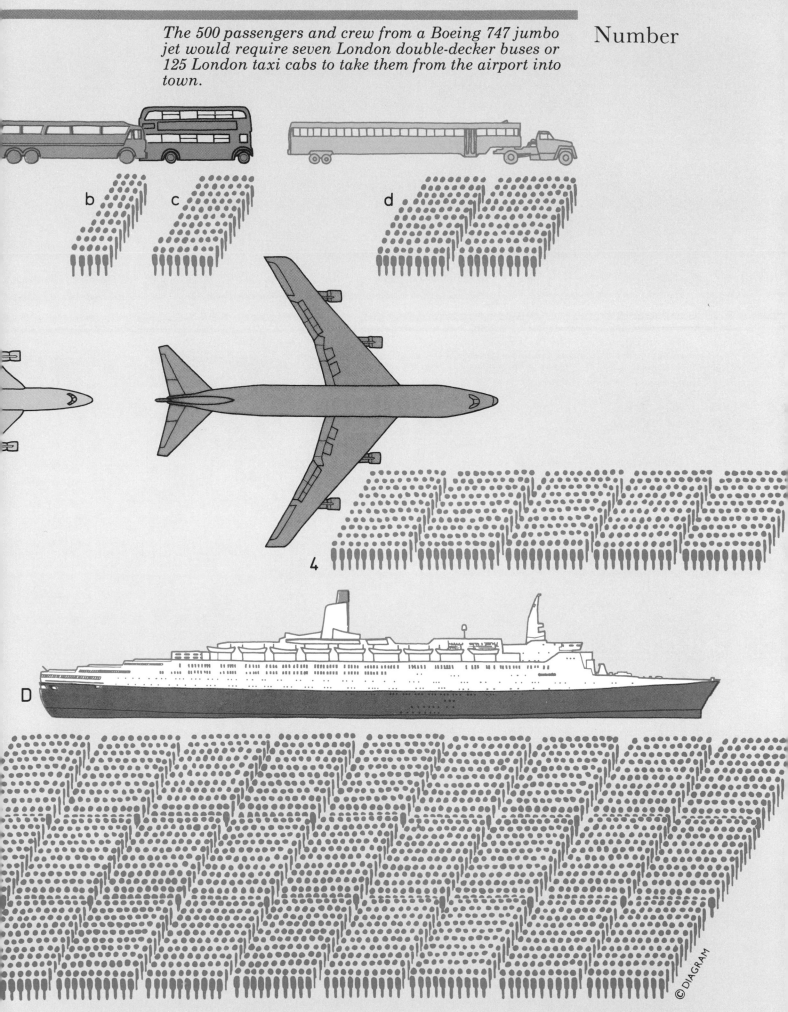

The 500 passengers and crew from a Boeing 747 jumbo jet would require seven London double-decker buses or 125 London taxi cabs to take them from the airport into town.

b

c

d

4

D

ODDS

The laws of probability are fundamental to any game of chance. Here we use three activities—tossing a coin, throwing dice, and poker—to demonstrate some basic principles. Odds can be expressed in two ways: for example, when a six-sided die is thrown, the chances of getting a specified number are 1 in 6, or 5 to 1.

Heads to win *right*
The diagram illustrates all the possible combinations of heads (**a**) and tails (**b**) for a coin tossed once, twice and three times in a row (with the situation after three throws summarized by the small circles below line 3). For example: the chance of heads in any one toss is 1 in 2; two heads in two tosses has the odds of 1 in 4 (1 in 2²); three heads in three tosses has the odds of 1 in 8 (1 in 2³). Following the same principle, ten heads in ten tosses has the odds of 1 in 1024 (1 in 2¹⁰).

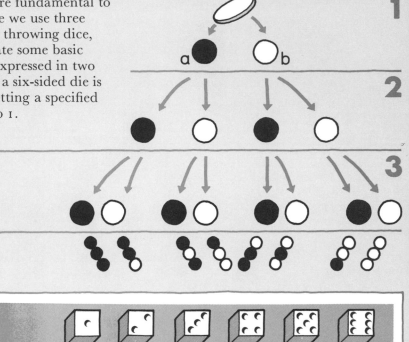

Dicing with fate
A six-sided die has an equal probability of landing on any of its six faces, therefore the probability of any specified number appearing at the top after a throw is 1 in 6. If two dice are thrown, the probability of a particular total appearing on the two faces varies with the number of combinations of the numerals 1 to 6 that will form that total. The table *right* shows all the possible totals from tossing two dice, and the ways in which they are formed. The chart *left* presents the same information classified by the total produced (**a**). From this information, odds against each total can be reckoned (**b**)—for instance, the chances of throwing a total of seven are 1 in 6 (1/6).

	a	b
	12	1/36
	11	1/18
	10	1/12
	9	1/9
	8	5/36
	7	1/6
	6	5/36
	5	1/9
	4	1/12
	3	1/18
	2	1/36

Odds for combinations in a four-child family are:
three of one sex and one of the other, 1 in 2;
two girls and two boys, 3 in 8;
three girls and one boy, 1 in 4;
three boys and one girl, 1 in 4;
four children of the same sex, 1 in 8;
four girls, 1 in 16; four boys, 1 in 16.

If we were to draw all the possible combinations of five cards to the size of those below, we would need 393,782 more pages. If we were to lay out all the cards of the possible hands, side by side, actual size, they would stretch over 484 miles (779km).

Five in the hand *right*
A person dealt five playing cards from a regular deck of 52 cards receives one combination out of a possible total of 2,598,960 (ignoring, as for poker, the order in which the different cards are dealt).

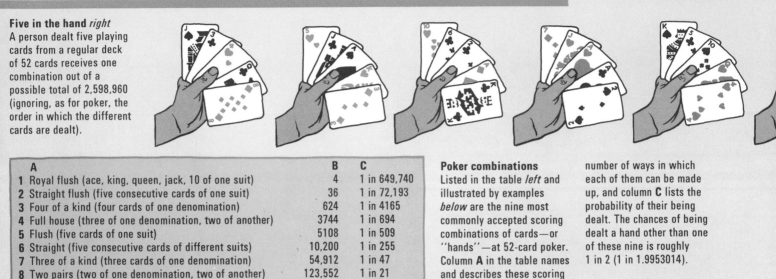

A		B	C
1 Royal flush (ace, king, queen, jack, 10 of one suit)		4	1 in 649,740
2 Straight flush (five consecutive cards of one suit)		36	1 in 72,193
3 Four of a kind (four cards of one denomination)		624	1 in 4165
4 Full house (three of one denomination, two of another)		3744	1 in 694
5 Flush (five cards of one suit)		5108	1 in 509
6 Straight (five consecutive cards of different suits)		10,200	1 in 255
7 Three of a kind (three cards of one denomination)		54,912	1 in 47
8 Two pairs (two of one denomination, two of another)		123,552	1 in 21
9 One pair (two of one denomination)		1,098,240	1 in 2.4

Poker combinations
Listed in the table *left* and illustrated by examples *below* are the nine most commonly accepted scoring combinations of cards—or "hands"—at 52-card poker. Column **A** in the table names and describes these scoring hands, column **B** lists the number of ways in which each of them can be made up, and column **C** lists the probability of their being dealt. The chances of being dealt a hand other than one of these nine is roughly 1 in 2 (1 in 1.9953014).

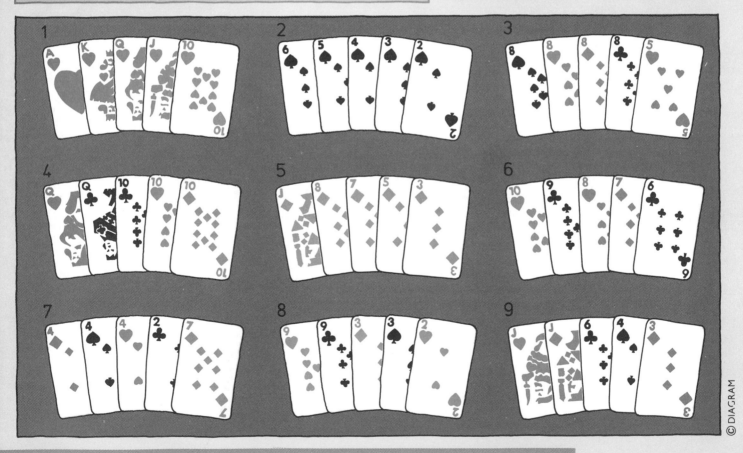

If 26 tiles, each bearing one letter of the alphabet, are placed face-down, the odds of the first three tiles to be picked up spelling, in the correct order, the word "end" are 1 in 15,600.

© DIAGRAM

INDEX

European countries are
shown fitted within the
vast area of the Indian
subcontinent in this
illustration from an
educational book published
in Calcutta in 1940.

This illustration from an
early 20th century edition
of Cassell's *Encyclopaedia*
gives a visual comparison
of the relative heights of
some famous buildings.

The comparative heights
of mountain peaks in
different ranges are shown
in this scale diagram from
*Systematische Bilder-
Gallerie* first published in
Germany in 1825–27.

INDEX

UNITS INDEX